Contents

Introduction

Finding out about structures can be fun. If you carry out the experiments and design projects suggested in this book, you will soon begin to realize how important it is to understand about structures.

Some tests and experiments will be easy to do, others will be harder. Work through them carefully and record what you find out. Sometimes you will need to look back in the book to refresh your memory. To help you do this, each chapter has 'key words' and ideas printed in bold type. Try and remember them and use them when you talk to your friends and teacher about your design ideas.

Sometimes you will need to discover things, so make guesses and see if you are right by reading on or testing and experimenting further.

Have some fun with your friends by playing the structures game in the middle of the book.

All these activities will help you understand more about the world of structures and how you can use this knowledge in design work.

STARTING DESIGN & TECHNOLOGY

STRUCTURES

ROGER BULL

Series editor: John Cave

CASSELL

Cassell Publishers Limited
Artillery House
Artillery Row
London SW1P 1RT

Copyright © Cassell 1989

First published 1989

ISBN 0-304-31646-6

Typeset by Flairplan Typesetting Ltd., Ware, Herts

Printed and bound in Great Britain by the Bath Press, Avon

1 What Is a Structure?

Man-made Structures

Is there a tower or bridge near your home or school?

Is it made in a similar way to those shown in the pictures?

What materials have been used to make it?

How are the parts joined together?

The Pompidou Centre in Paris is an 'inside-out' building.

The Eiffel Tower. It is made of iron and is over 300 m tall.

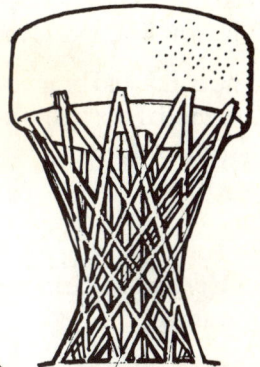

A water tower built of reinforced concrete.

The steel structure is on the outside of the Pompidou Centre.

The first bridge to be built of iron at Ironbridge, Shropshire, was completed in 1775.

The Munich Olympic Stadium has an acrylic (plastic) sheet roof held up by steel cables and steel tubes called pylons.

A barn structure built of wood in about 1800.

What Is a Structure?

Structures in the Natural World

Trees grow tall and spread their branches so that their leaves are open to the light.
The trunk supports the branches and leaves.

The trunk and branches must be able to resist the wind if the tree is not to get damaged or blown over.

The trunk supports the branches and leaves.

The trunk and branches deflect (move) when resisting strong winds.

Here are some more examples of structures in nature.

The wings of a bat are stretched out by bone and cartilage.

The skeleton of a horse.

The micro-structure of bone.

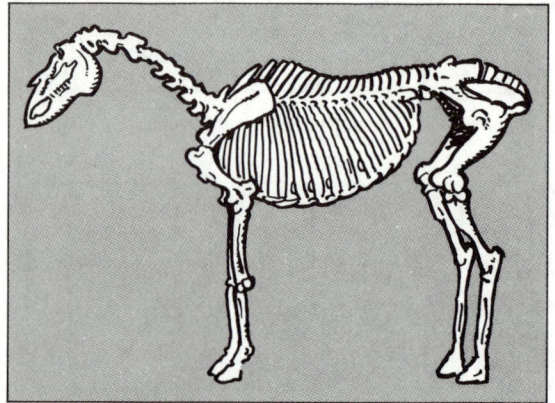

Honeycomb structure made by bees.

A spider's web

Fact File

- Some natural structures, like the bracken plant, can grow as tall as 4.5 metres.
- The world's tallest trees grow to a height of 415 metres and belong to the Douglas fir family.

Structures in Transport

The Sopwith Camel (1917), a fighter plane from the First World War. The wings are covered in fabric and held apart by wooden 'spars' and tied together with steel wire.

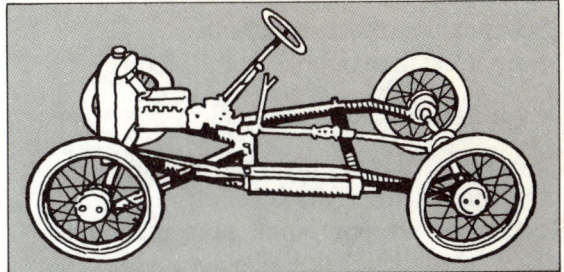

Different types of **transport** are good examples of man-made structures. Transport is a way of carrying people or goods around.

You can see that modern transport is very different from that of the past. New materials and methods of construction have enabled the designer to develop new ideas for transport with different purposes.

Take one structure from each page and for each one make a list of its **functions**. These are the things it has to do.

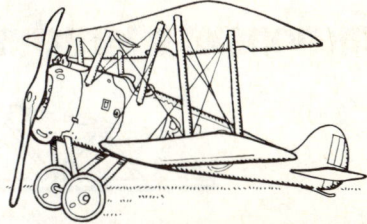

The chassis of an Austin Seven (1922). A chassis is a frame on which the body is bolted.

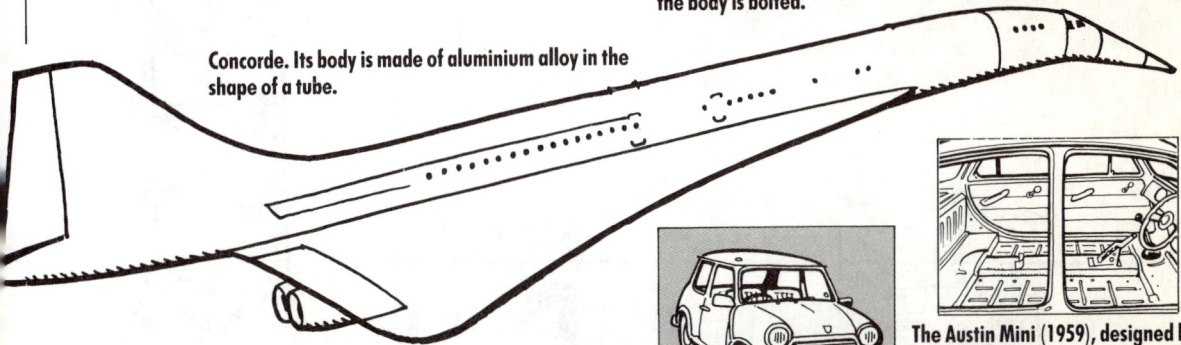

Concorde. Its body is made of aluminium alloy in the shape of a tube.

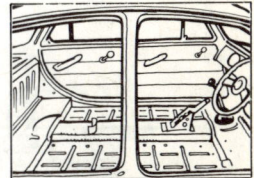

The Austin Mini (1959), designed by Sir Alec Issigonis. It did not have a chassis; its floor was pressed out of a sheet of steel.

The modern racing car is made with a body which has only a single skin. This is called monocoque construction.

The frame of a Moulton bicycle is different from that of a BMX bike.

Here is a list of the functions of a bridge:
- It must support the weight of all the traffic crossing it at any one time.
- It must hold up its own weight.
- It must resist the wind and not get blown down.
- It must resist the flow of the river and not get washed away.

Can you think of any more?

What Is a Structure?

The Functions of a Structure

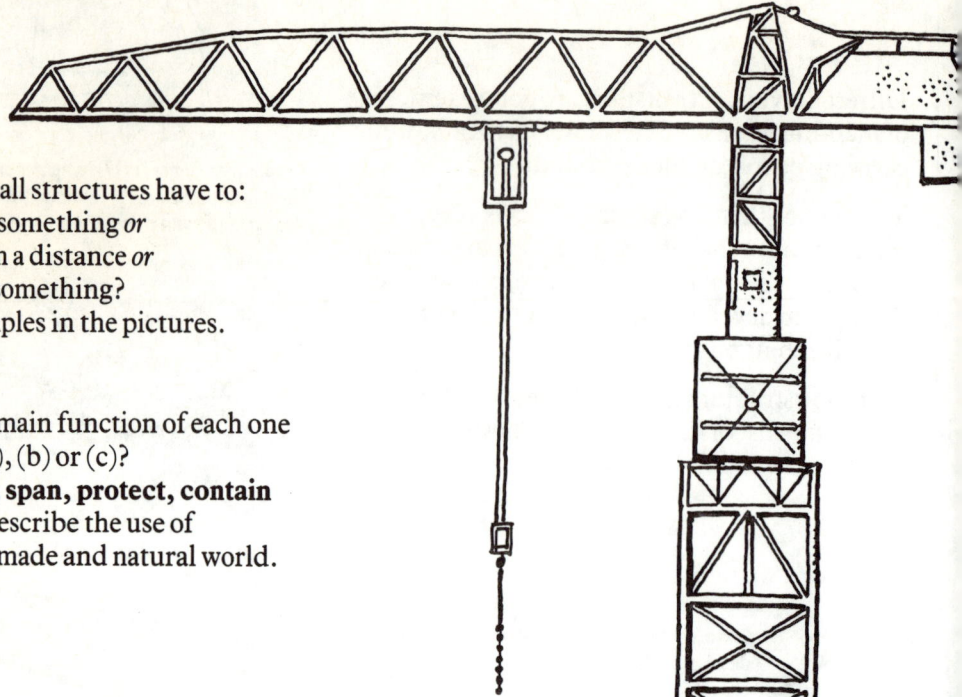

Have you noticed that all structures have to:
(a) hold up or support something *or*
(b) reach across or span a distance *or*
(c) contain or protect something?
Look back at the examples in the pictures.

Do you agree?

Do you think that the main function of each one
is the same as either (a), (b) or (c)?
Remember: Support, span, protect, contain
are key words which describe the use of
structures in the man-made and natural world.

A North Sea oil platform is supported by a steel and
concrete structure.

A pet's travel cage contains and protects.

A tower crane 'reaches' across. A counter-weight balances the load.

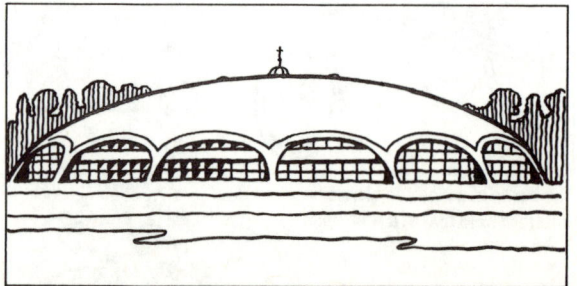

The roof spans a large area, leaving plenty of space below.

Some Functions Are Not So Obvious

The designer of a chair must make sure it is comfortable, but it must also be able to support your weight, even when you tip backwards on it.

The joints and legs should not break.

The designer of a container for a fizzy drink must make sure that it can withstand the gas pressure inside and the knocks it will get when carried about. An empty lemonade can will support you if you stand on it, but it will collapse if you squeeze its side.
Do you know why?

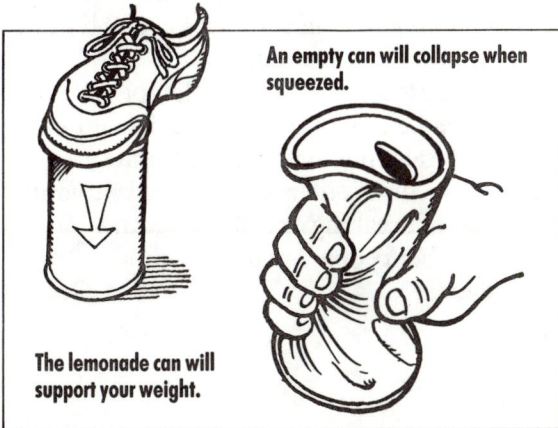

An empty can will collapse when squeezed.

The lemonade can will support your weight.

You may never have thought of a drink can as a structure, but it does have to contain a liquid as well as support the weight of other cans stacked on top of it. How many objects can you think of that stack?
The pictures show some of them. Can you think of any more? What do you notice about the shape of the objects? How are they designed to resist the weight of things above them?

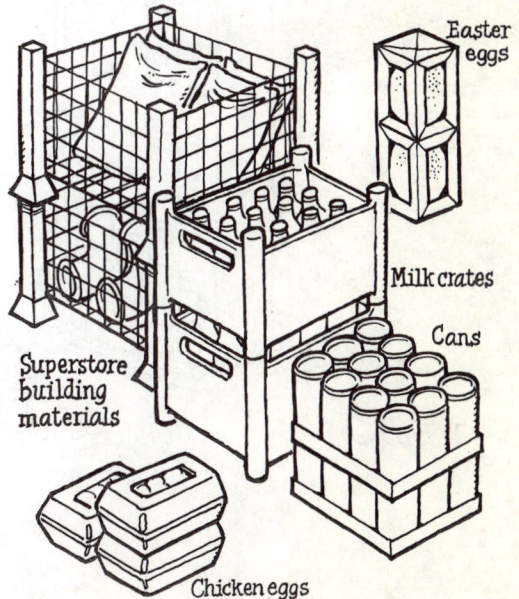

Easter eggs

Milk crates

Cans

Superstore building materials

Chicken eggs

What Is a Structure?

What Must a Designer of Structures Find out About?

A person who builds a good structure, one that carries out all its functions successfully, will need to have thought and found out about:
- *Materials* and what happens to them when used in a structure.
- *Things that affect structures*, like
  - (a) the wind and rain and the effect of different types of weather on it;
  - (b) the weight of things they carry or support;
  - (c) their own size and weight.

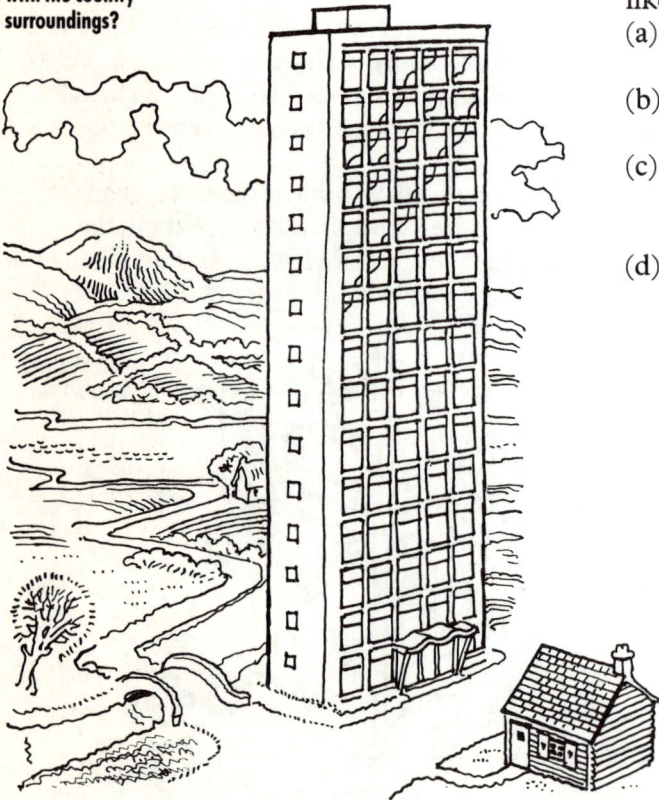

How many clothes will it support?

How much will it carry?

Does the tower block fit in with the country surroundings?

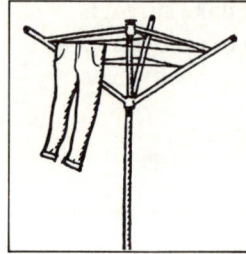

- *The limits, or constraints, of a design problem,* like
  - (a) *cost*, or how much the customer is willing to pay;
  - (b) *appearance*, or how well it looks and fits in with the surroundings;
  - (c) *ability to construct*, or whether people have the skill, knowledge and tools to make and put together a structure;
  - (d) *safety*, or whether it can be used without risk of breaking or collapsing, and whether people can use it without danger.

How Will the Structure Be Assembled?

Just imagine if you had designed and built a
building or a bridge and then you were unable
to get the last piece in place!
As you work through the problems and tasks in
the following pages, you will see how important
all these things are.

The designers of these bridges have worked out how they should be
made and in what order they should be assembled.

2 Structures and Forces

Structures Are Affected by Forces

Forces can:
- get things moving
- stop things moving
- change the direction of movement
- change the shape of things

We can see the result of forces acting on things.

The pictures show some forces acting on things.

Can you describe the force and say what will happen?

A sailing dinghy

A broken-down car

Dodgem cars

A horse and trap

Traction engine tug-of-war

Fact File

- The Eiffel Tower, built in 1889, is an iron structure that sways up to 12.7 cm in the wind.
- The wind is a force that moves the tower, but it does not buckle or break because it is designed to resist that force.

The Force of Gravity

Gravity is an important force and pulls things to the earth.

Here is an average-sized apple. If it rolled past the edge of the table, it would be pulled to the ground by the force of gravity. If it were hung from a piece of elastic, the elastic would stretch. We see the result of the force because the elastic stretches.

Gravity is pulling the apple with a force that is called the weight of the apple.

Weight is a force. It is measured in **newtons**. If we were to add more average-sized apples to the elastic, one by one, it would stretch an equal amount each time.

The more apples there are, the greater the pull to the earth.

The pull of gravity

An average-sized apple is pulled to the ground.

Elastic band stretches

Elastic band stretches further each time a friend comes along

The more 'friends' there are, the greater the pull to earth. The force stretching the band is greater and is measured in newtons.

You will know that food and materials are often measured in **kilograms.** Scales enable us to measure them so that we know we have the right amount.

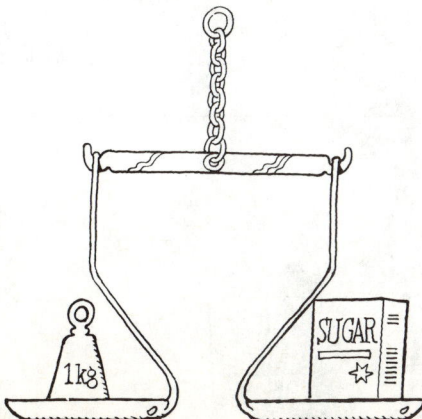

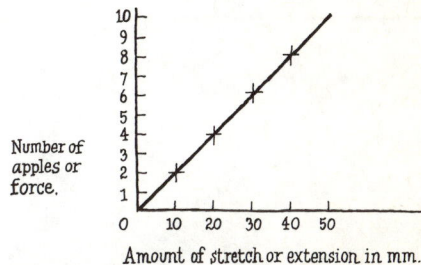

Number of apples or force.

Amount of stretch or extension in mm.

This graph is a record of how much the elastic band stretched.

Food was always measured by balancing it against an amount of metal called a **mass**. This is a standard measure used throughout most of the world. Today scales are first set up and tested with a standard mass.

The mass of an object is a measure of how much matter is in it, e.g. the amount of sugar in candy floss or the amount of plastic in a swimming float.

1kg   SUGAR

Structures and Loads

Fact File

- On the earth, a mass of 98.1 g will be pulled to the ground with a force of 1 newton (1 N). The average-sized apple has a mass of about 98.1 g, so it will also be pulled by the same amount of force, which is about 1 N.
- The pull of gravity gets less the further you are away from the earth. If you were lucky enough to fly into outer space, you would find the apple with the same mass of 98.1 g would be pulled down with much less of a force.

The nearer the mass is to the earth's surface, the greater the pulling force.

We say that a structure is loaded when a force acts on it.
What type of loads do the structures in the pictures carry?

A bridge

A tipper truck

A shopping trolley

Structures and Loads

Have you named things like cars, lorries, rubble, shopping goods, gravel and sand?

You are correct, but you should remember that forces are calculated in newtons.

We can make approximate calculations of the force of gravity by saying that a mass of 100 g acts downwards with a force of 1 N.

If a shopping trolley is carrying a mass of 50 kg, then a force of approximately 500 N is acting down on it. This is found by doing this calculation.

$$\text{If a mass of } 100\,\text{g} = \text{a force of } 1\,\text{N}$$
$$\text{and } 50\,\text{kg} = 50{,}000\,\text{g},$$
$$\text{then the force} = 50{,}000 \div 100$$
$$= 500\,\text{N}.$$

Tests or calculations are often carried out to find how a structure will behave under load or what it can carry safely.

Fact File

Civil engineering structures like dams and tunnels are enormously strong and support massive loads. In the USSR there is a dam on the river Yenisey capable of holding back 31,300 million cubic metres of water, which is estimated to be a load of 8,000,000 tonnes. Since the strongest human being has been known to support a load of 2,844 kg (2.84 tonnes), it follows that it would take 2,816,901 of the strongest people to hold back that amount of water. Of course this is not possible, since their hands would leak!

3 Shape and Stiffness

Bending under Load

Experiment 1

You will need:
- 1 piece of thick card or thin wood
- support blocks
- some weights

Place a weight of 1 N on the middle of the wood or card **beam,** or hang a 1-N weight from it. You have applied a force of 1 N. The force is acting downward. What has happened? Continue to load the beam. You will notice it is **deflecting** (moving). Record how far it deflects under a certain load.

Experiment 2

Take away the card or wooden beam and replace it with a piece of steel about 3 mm thick. Gradually load the metal beam.
How much does it deflect?
Will it deflect at all?
Record the deflection for each loading. Compare the results of this experiment with those from Experiment 1.
What can you say about the movement of the wooden or card beam compared with the metal one? Is it greater or smaller? Does this mean that wood or card is more flexible than the metal, or stiffer?
Remember: Stiffness is the ability of a structure to withstand being bent, twisted, stretched or changed in shape.

Wooden beam. One person, no deflection.

Wooden beam. Two people, deflection.

Metal beam. Two people, very little deflection.

Bending under Load

You have seen a card or wooden beam deflect under a load and a metal beam remain stiff. Some materials are stiffer than others.

Do you think it is a bad thing that beams deflect?

How far should they be allowed to deflect? Think back to the way the Eiffel Tower moves because of the force of the wind.

Where do you find beams being used? A beam **spans** a space and is usually supported at both ends. The simplest type of bridge is a beam.

Fact File

- Structures like oil tankers are built to bend and flex in the water. If they were rigid structures, they would 'break their backs'.
- Oil tankers can be as long as 450 metres. Just think of the construction and launching problems!

A diving board is a special type of beam. It needs to deflect the right amount to give the diver the right amount of spring.

Carrying a Load

A transport company has one trailer which it must use to carry a very heavy **load** from Deepdale to Highhill. (You can build your own rough terrain from plywood or hardboard.) When the load is placed on the trailer, the surface is deflected. It **bends** almost to the ground. When the load reaches Highhill, the bottom of the trailer scrapes the road surface. This is a major problem. Can you redesign the trailer and overcome the problem?

The trailer can be modelled from 3mm thick card or balsa wood.

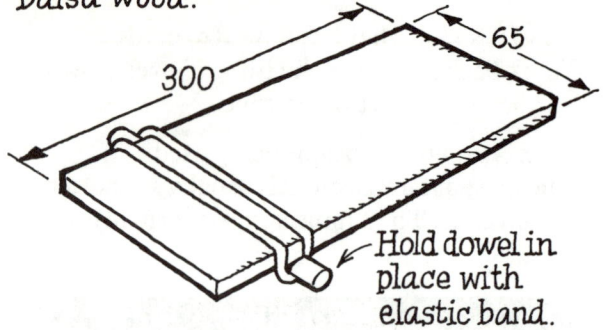

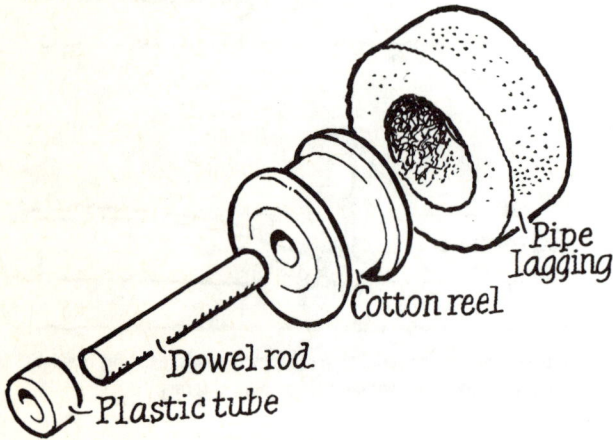

300

65

Hold dowel in place with elastic band.

Pipe
Lagging

Cotton reel

Dowel rod

Plastic tube

Wheels can be made from cotton reels and pipe lagging held in place with plastic tube.

Carrying a Load

How can you improve the **stiffness** of the trailer?

Look at the pictures of lorries and trailers that are designed to carry loads. Carry out your own experiments and tests. Notice that the trailer surface is a beam supported at both ends by the **axles** and wheels.

Remember: The trailer must not bend, but it must remain as light in weight as possible. If you increase the load which the engine has to pull, the lorry might not make it up the hill! Here are some suggestions for starting-points in your experiments and tests.
- how many supports?
- where should they be positioned?
- how many axles are there on the trailers?
Remember to record your results.

Fact File
- Loads of over 90 metres in length have been transported in Britain.
- How many axles should there be on the trailers to carry them?

Improving Stiffness

What happens if you add more axles and wheels to your trailer? This is the same as saying 'add more support to the beam'.

What happens if you move the axles closer together? This is the same as saying 'share the support of the beam'.

What special problems arise when you do these things? You will increase the overall weight of the vehicle, but how will it affect the steering? Construct your model and find out.

Let us keep just two supports for the trailer. The trailer must be a certain length because the load is a particular size. We are unable to alter the length or the number of axles. What else can we do to reduce bending in the trailer?

More axles give more support.

Carry out some tests.
Notice that the trailers have beams that travel their length. One trailer has a beam with 'diamond holes'. What is the reason for this?

If you were to see their end-shape (cross-section) they would look like this.

This gives an idea about how to stiffen the trailer.

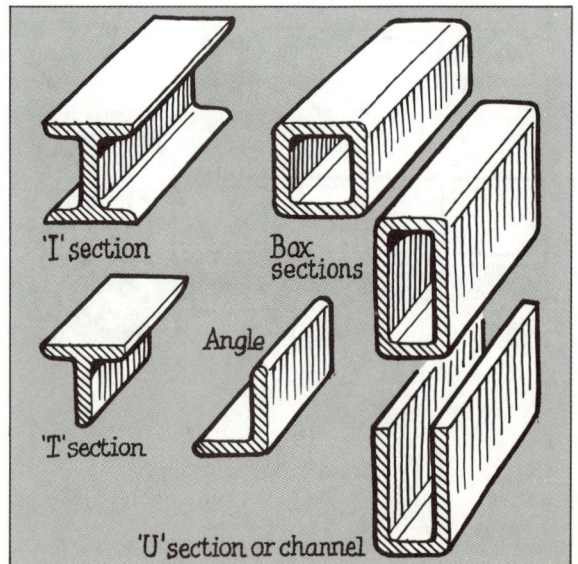

Cross-sections of beams

'I' section   Box sections

'T' section   Angle

'U' section or channel

Improving Stiffness

Make and test these improvements to your trailer.

Glue wood along the edge of the trailer. This is similar to folding over the edge of card or metal. What can you say about the stiffness of the trailer now?

Fold a piece of paper like this. Sandwich it between two pieces of 1.5 mm card. Test for deflection. What other in-fill shapes might we use?

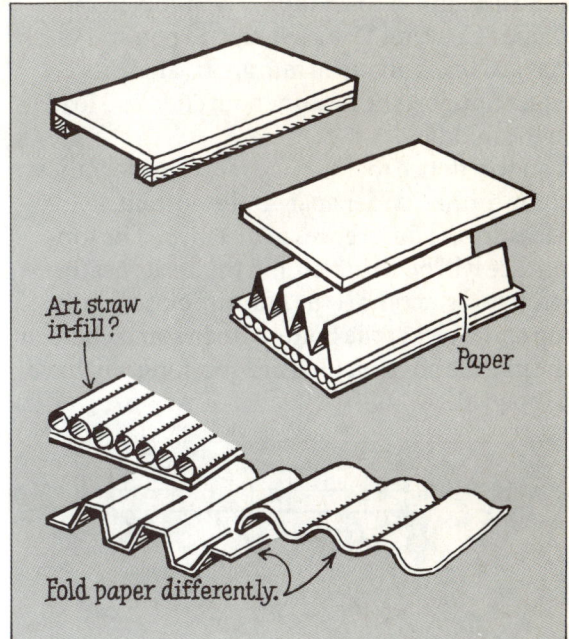

Art straw in-fill?

Paper

Fold paper differently.

Make different-shaped beams from card. Here are some ways of making beams in card. Support them from the ends and load them.

How much loading will each beam take? What happens when failure occurs? Has the girder: (a) bent, (b) buckled, (c) stretched or (d) twisted?

You will now see that the **shape** of a beam is very important. So the solution to the trailer problem will have something to do with the shape of a beam and the position of the axles.

Forces that tend to bend can be resisted by altering beam shapes and the number of supports. Test your trailer out now. How much weight can it support?

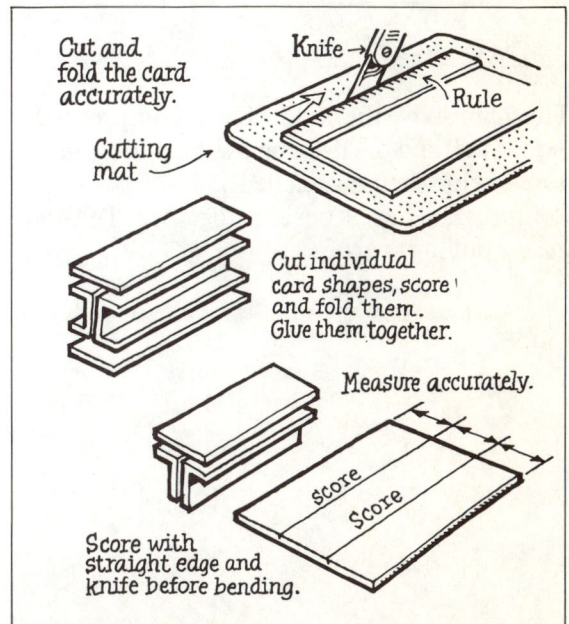

Cut and fold the card accurately.

Knife

Rule

Cutting mat

Cut individual card shapes, score and fold them. Glue them together.

Measure accurately.

Score

Score

Score with straight edge and knife before bending.

4 Compression and Tension

Squashing and Stretching

Find a length of foam rubber, roughly in the shape of a beam. Use a felt-tipped pen and mark vertical lines, about 20 mm apart, along its edge. Support the beam at both ends and load it in the middle.

You will notice that the lines are close together at the top and wider apart at the bottom. Measure the differences in distance. The top surface will be wrinkled and the underneath stretched smooth. If the lines are closer at the top and wider at the bottom, then somewhere in the middle no stretching or squashing will have happened.

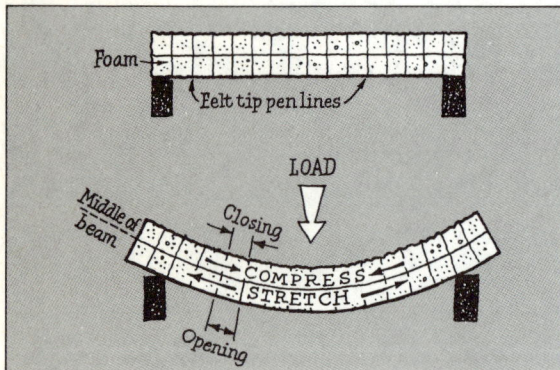

The foam above the middle line is **compressed** (squashed) and the foam below the middle line is in **tension** (being stretched).

Compression means pushing together. Tension means pulling apart.

Look at the following diagrams. Notice how the labelled parts are being pushed together or pulled apart.

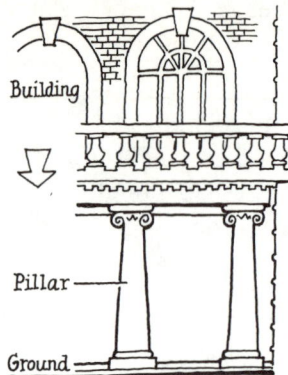

A two-storey building

Child on a swing

Body-building

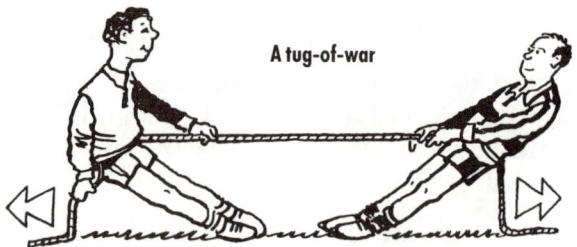

A tug-of-war

The arrows show the direction of outside or **external forces**. These forces put objects under compression or tension.

When a structure is put under load, it must be able to resist it. When a weight-lifter holds up a load, he must push up by the same amount as the weight of the bar. What would happen if he did not?

Internal and External Forces

If one force is unequal to another, something must happen. Look back to Chapter 2 to decide what this will be.

Experiment 3

Pull on a piece of elastic. The external force you exert will stretch it. Now relax slightly. Reduce the external force. The band will pull your hands inward. Release a little more. Your hands will go back further. There is a force within the band which pulls it back to its original length. This force is the **internal force**, and it acts against the external force.

The internal force must be equal to the external force if the object is to remain in **equilibrium**. In other words, if the band does not stretch any more, then the internal forces are equal to the external forces.

Forces must balance in a structure or the structure will either bend, buckle or break, stretch or compress (fail) – or something else will happen.

If the force at one end of a see-saw is not balanced by the same force at the other end, it will tip.

Carry out the test and you will find that the positions of the forces are important.

Different forces can be balanced by changing their positions. When balanced, the see-saw is in equilibrium.

Can you explain how a tower-crane balances?

TENSION
PULL — INTERNAL FORCE — PULL

COMPRESSION
PUSH — INTERNAL FORCE — PUSH

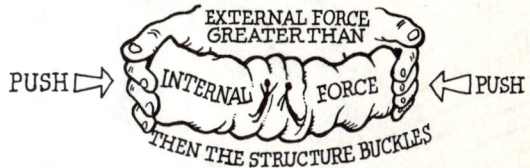

EXTERNAL FORCE GREATER THAN
PUSH — INTERNAL FORCE — PUSH
THEN THE STRUCTURE BUCKLES

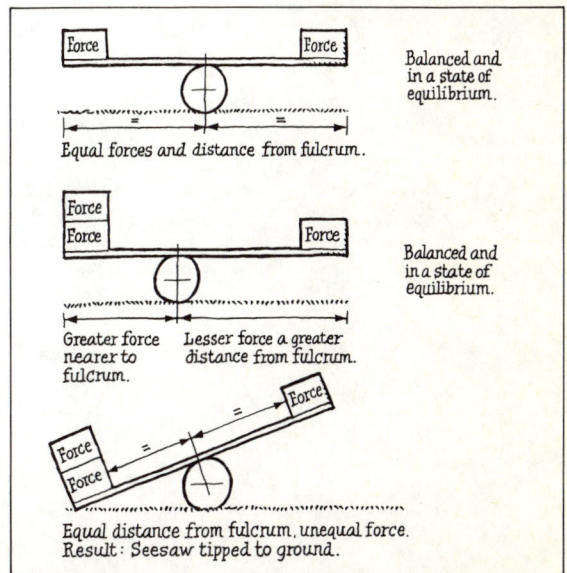

Force — Force
Balanced and in a state of equilibrium.
Equal forces and distance from fulcrum.

Force / Force — Force
Balanced and in a state of equilibrium.
Greater force nearer to fulcrum. Lesser force a greater distance from fulcrum.

Force — Force / Force
Equal distance from fulcrum, unequal force.
Result: Seesaw tipped to ground.

Compression and Tension

Forces and Materials

Experiment 4: Testing Materials

You can carry out some simple tests to see how certain materials behave when they are compressed or put under tension. These are the most common forces that a material in a structure will be subject to.

The pictures show equipment that could be made for you so that you can do your tests.

This is a tensile testing machine.

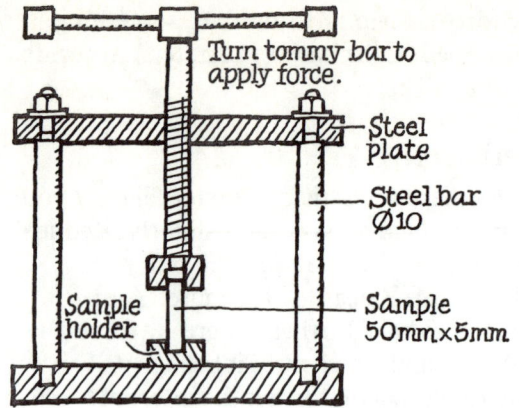

Turn tommy bar to apply force.

Steel plate

Steel bar Ø10

Sample holder

Sample 50mm×5mm

This is a machine for testing compression.

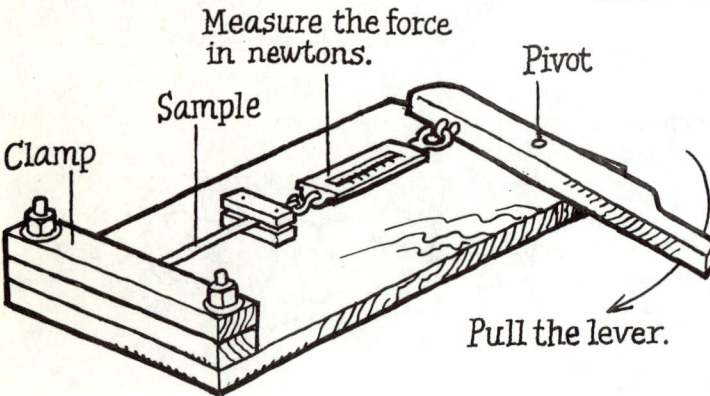

Measure the force in newtons.

Pivot

Sample

Clamp

Pull the lever.

Use the following words to describe what happened:

twist, fold, bend, powder, stretch, fracture, flex, buckle, break, explode, splinter.

Remember: You are comparing the behaviour of one material with another.

Here are some suggestions of materials for testing.

Compression	Tension
wooden dowel rod	soft iron binding wire
aluminium	fishing line (with
chalk	different breaking strains)
(*Note:* samples should be no more than 5 mm in diameter)	elastic band aluminium wire thread

Recording Your Results

The tensile machine enables you to measure the pulling force you apply and what happens at that stage of the test. What you see happening is important but also describe what you feel happening through the lever.

Name of material	Tensile force applied	Description of material's behaviour
Elastic band	1 N	Becomes taut
	2 N	Stretches 3 mm
	25 N	Snaps suddenly

Torque wrench measures the turning force.

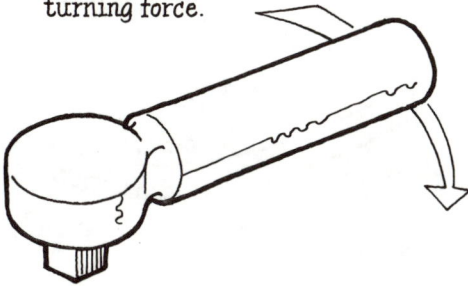

Torque wrench socket with slots cut so that it locates over tommy bar.

In the compression test, count the number of turns you make with the tommy bar, or your teacher could adapt a torque-wrench socket so that the turning-force applied can be measured in newton/metre. Material scientists use special and more accurate types of machine and often record their results in the form of graphs. They are concerned with finding materials which are suitable for use in structural design.

Compression and Tension

Strength

We say materials have **strength**. Strength is how much force a material can stand before it **breaks**.

Experiment 5

Bend a piece of chalk. It snaps suddenly.
Bend a piece of dowel rod. It splinters, then breaks.
Bend a piece of mild steel. It does not snap but will keep the bend.
Bend a piece of rubber. It returns to its original shape, but it might also tear and then break.

Which material is the strongest when subject to bending forces? (*Note:* To make a comparison, each sample should be the same size.)

Now test the materials in another way – compress them. Which material is the strongest? Look at your test results to find out.

Fact File

You would have to build a chimney or brick structure over 6.5 km high before there was a risk that the bottom bricks would be crushed by the weight of those above.

CHALK — SNAP!

DOWEL — splinters, breaks

STEEL — bends

RUBBER — tears and breaks

Stress

The amount of pushing together (compression) or pulling apart (tension) before breaking tells us how strong a material is.

If a material is bigger in cross-section (that is, its end-shape is bigger in area) it will always be stronger because more square millimetres are sharing the force.

To compare one material with another, we find out how much force each material is able to resist in 1 sq mm. This tells us how much **stress** the material can stand.

1 sq. mm shares the pulling force with 15 more sq.mm.

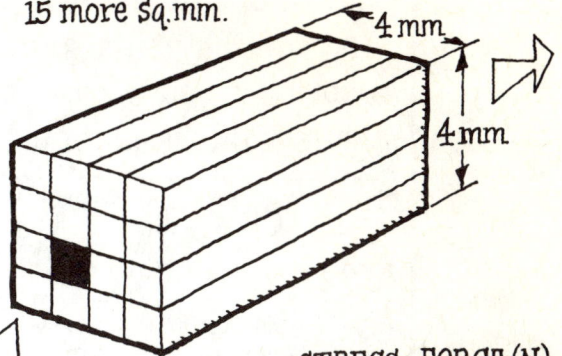

4 mm

4 mm

$$STRESS = \frac{FORCE\ (N)}{16\ sq.mm}$$

Small cross-section.

Big cross-section.

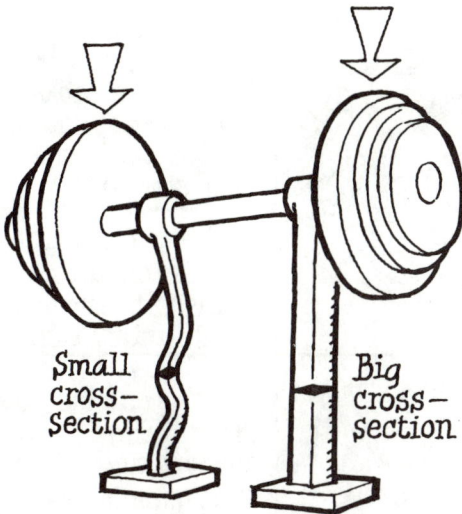

When more square millimetres share the load, the support becomes stronger.

Materials are also tested to see how **tough**, **elastic** and **hard** they are.

Toughness tells how much force a material can absorb before it breaks.

Elasticity tells how well a material goes back to its original shape after being stretched or compressed.

Hardness tells how well a material resists being scratched.

Further tests and calculations are carried out to find a material's other qualities so that the designer knows it can resist all the forces it will be subjected to in a structure.

NEWTON · THE AVERAGE–SIZED APPLE

Can you solve all the problems on your journey before 'Newton', the average-sized apple, falls to the ground? How far can you get?

You can play this game no matter where you are in the book. As you become more experienced in structures work you should be able to travel further in your journey.

① How can I get Newton down without damaging him?

② How can I protect him so he doesn't get bruised?

③ PLEASE SHUT THE GATE

The gate is broken. What shape could I make a new one so it is strong ?

⑥ It's getting very hard work. What could be stopping me? How could my friends help me keep moving?

⑦ Oh no! The bike is out of control. CRASH! I can't ride it now. There are several signs of failure in the bike's structure. What could they be?

CRASH

⑧ I'll place all the apples in crates and transport them to the cider factory. They are very heavy. The stack wobbles and the crates are bulging and breaking. What materials and shapes should I use to make the crates stronger and safer when stacked ?

NEWTON'S AT THE START OF HIS JOURNEY !

MOVE **ALONG** FOR EVERY SOLUTION.
⟹ *IT WILL BE DIFFICULT*

	Ground level	1	2	3	4	5	6
1							
2							
3							
4							
5							
6							

A STRUCTURES TECHNOLOGY GAME

The rules are simple ~ Throw a dice to see how many solutions you must suggest. $\boxed{\cdot}$ = 1 solution, $\boxed{\vdots}$ = 5 solutions. On the bottom grid, move forward one square for each sensible solution you suggest. If you fail to suggest the correct number of solutions, go down a level.

④

How can I get across?

⑤ A bicycle would be quicker. Newton is on the carrier. Before you ride it, you must describe things about it that are to do with forces and structures.

STORE

The winch beam is bending.

Don't climb this ladder.

⑨

⑩ What parts of the building are in tension and compression?

SNAP

Who ever made this bridge? What can I do to make the road surface stay level?

PLATFORM

⑫ Tip into hopper

PRESS ROOM

How can I apply a force and squeeze the apples between the plates?

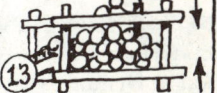

Phew! After all that effort I must have a rest. I hope the bed doesn't collapse!

⑪ Can you suggest ways of improving each broken or unsafe item, so the factory is safe again?

The trailer has tipped up.

⑬ What joints and shapes should I use in the frame of the press?

SIX!

MOVE **DOWN** ONE LEVEL ONLY IF YOU FAIL TO GET THE NUMBER OF SOLUTIONS.

SOMETIMES TO THINK OF THE NUMBER OF SOLUTIONS SHOWN ON THE DICE.

							Did you make it?
7	8	9	10	11	12	13	

5 Dynamic and Static Forces

Still and Moving Forces

How well did you sleep last night? A strange question you might think, but let us consider things a little more carefully. A bed is a structure. It supports you while you sleep. A bed is basically a framework which supports a large flat surface. Normally this surface is made up of springs, which support the mattress on which you lie.

The bed needs to support your weight, but, unfortunately for the bed, you enjoy trampolining! Bouncing on the bed is great fun, so long as your parents do not catch you doing it and so long as it does not collapse when you are doing it. Luckily, most beds are capable of supporting you both when you are asleep and still, or **static**, as well as when you are bouncing up and down or acting in a **dynamic** way.

Static forces and dynamic forces have to be resisted by all structures; otherwise they will fail.

Dynamic forces are **destructive**. They break things up. Placing weights on to a surface produces a static force; allowing weights to fall on to a surface produces a dynamic force.

Fact File

The largest bed in Britain is now kept at the Victoria and Albert Museum. 'The Great Bed of Ware' is 10 ft 8½ inches wide × 11 ft 1 inch long × 8 ft 9 inches tall (3.26 metres × 3.38 metres × 2.67 metres).
I wonder how many people could bounce on it?

Static

Dynamic

Destructive

Failure

When you bounce on your bed, the frame might buckle, twist, bend, kink or break. All these words describe the **failure** of the bed. Many structures, including beds, are capable of resisting far greater forces than would be expected. Up to four or five times more. Why do you think this is?

Look at the list of items below. For each one, write down whether or not it resists static or dynamic forces or *both*. Think very carefully!

Rotary washing line
Walls of a house
Horizontal bar in gymnastics
A beam in the gym
Cricket bat
Car

Ask yourself this question: Are the forces moving or 'resting'?

For each item, give an example of when **structural failure** might be said to have occurred.

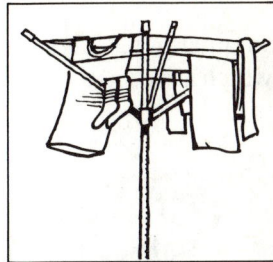

Rotary washing line

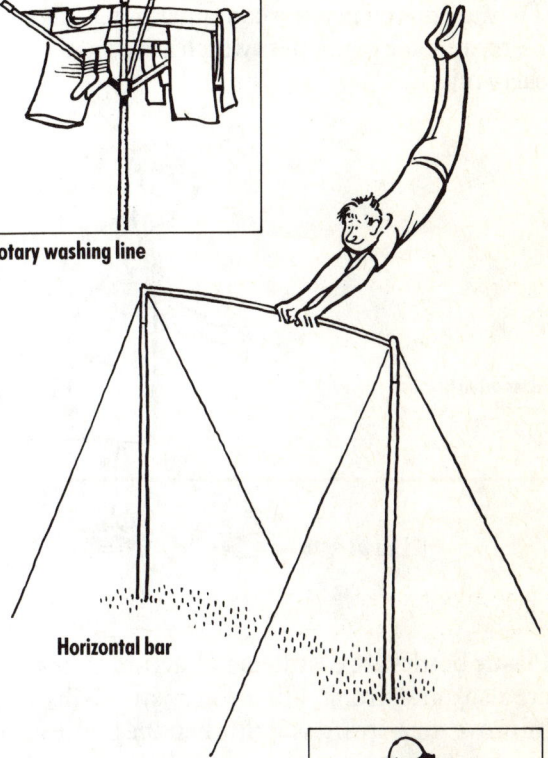

Horizontal bar

Walls of a house   **A car**

A cricket bat

31

Packaging and Containers

Packaging protects food and drink from static and dynamic forces when it is being carried about.

How are foods like strawberries, eggs, salt and tea protected, and kept fresh and clean?

How are liquids like milk, wine and oil carried about?

They are placed inside a container or package. Here are some containers which are used to carry milk.

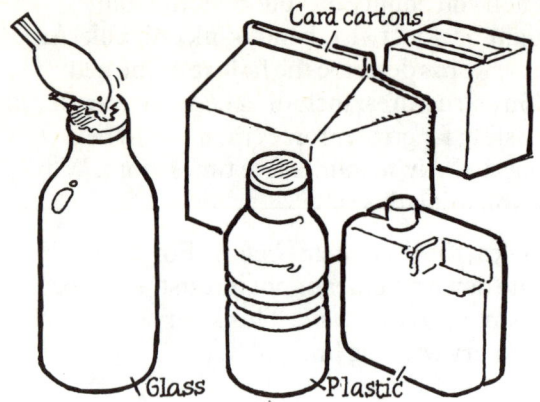

Card cartons

Glass   Plastic

Glass often breaks.

Dangerous

Why do you think a number of different methods of containing the milk are used? Are some methods better than others? Here are some advantages and disadvantages of each. If you drop the milk bottle, it is likely to break, but glass is very easily cleaned and sterilized.

Unfortunately, bottles must be returned to be refilled. However, they do not leak because they have no seams or joins. They are stiff when handled and pour well.

Plastic bottles are easy to manufacture. They are also quite strong, but not very stiff. Ribs improve their stiffness, but when the bottle is partly empty it can sometimes collapse or kink. They are disposable, so no collection service is needed. When dropped, plastic bottles are likely to split.

Annoying and inconvenient but less dangerous.

Plastic may split.

Packaging and Containers

Any number of milk cartons can be stacked together. Their flat, folded tops make this possible, although some cartons are produced with different-shaped tops. The cartons are made from waxed card, which is the cheapest of the materials used in milk container design. If dropped, a carton may rip or split along a join, but it is less likely to do so than any other container. It is disposable and it is easy to print on when in the form of a flat piece of card.

Flap top for pouring is turned down.

The straight lines show where bends are made.

Gluing the joins together is not always successful, so milk cartons can often leak. Their greatest disadvantage is that they are difficult to open. People with weak fingers and arthritic hands find opening a carton almost impossible. It is also annoying to have to cut a flap off before pouring.

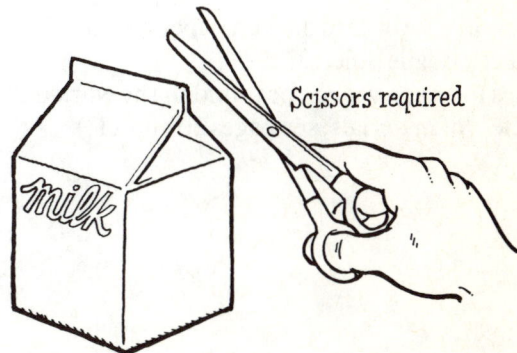

Scissors required

Consumer Research

Project 1: Packaging Survey

List a number of ways of holding, containing and protecting a product like wine, soap (solid and liquid), eggs (chicken or chocolate), engine oil.

Write down some questions about how good or bad and how useful the container or package is. Now carry out a research survey by asking friends and their parents for answers to your questions. Here are some sample questions:

How many eggs do you buy a week?
How often do you find a broken one on returning home?
Is the type of packaging easy to open?
Does it do its job well?

Remember: All these containers are structures. While they can resist some forces, they cannot resist others.

We have already discovered that dynamic forces are destructive: for example, a milk bottle smashes when dropped. Static forces do not present such a problem: milk cartons, for instance, can support many more stacked on top of them.

How is a piece of card or even paper able to protect a fragile object?
(Clues to an answer can be found in the work on stiffness of materials; see pages 20 and 21.)

HOW EASY IS IT TO OPEN?

What questions will you ask? What product will you choose?

Changing the Shape of Materials

We must change the shape of the paper. Wide, flat surfaces will buckle and fold under load. It is easy to crumple a piece of paper.

Now roll it up and hold it together with an elastic band. It is much harder to crush.

Place a series of rolled-up pieces of paper on a baseboard. Then put a board on top of them, and carefully step on it.
Now that the shape has been altered, the paper is much stronger.

Solid materials are the best at supporting loads, but in structures lightness is important as well.

The honeycomb is a hexagonal structural shape in which strength is gained by the close fit of every cell.

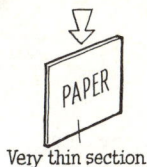

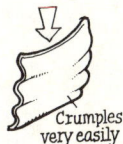

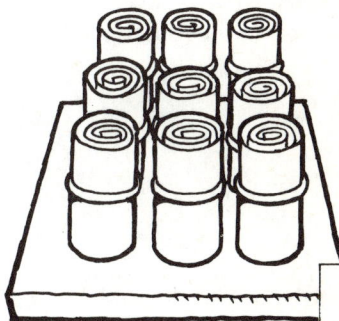

PAPER

Very thin section

Crumples very easily

Elastic band

Honeycomb

Solid, strong

Square to be lighter and still strong

Fact File

- The world's tallest man-made structure is a radio mast in Poland. It is 646.38 metres tall.
- It is made of tubular steel. Why is this so important?

Folding Card Projects – Packaging

Project 2

Using just one piece of A4 card and stick-type glue, make a package that will offer the best possible protection for a hollow chocolate egg which is the size of a chicken egg.

Project 3

Using two pieces of A4 card, make a package for a presentation mug. The package should protect the mug, at the same time as allowing the mug to be seen through its sides.

Remember: Score the card where folds are to be made and plan the shape of the folded form.

What shape should the outer box be?

Egg carton support   Could the paper act as a spring?

How will the egg be supported and protected?

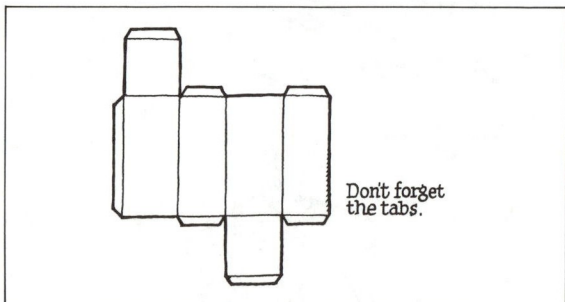

Don't forget the tabs.

Where must the paper or card be folded? What printing and decoration will you apply?

The mug problem requires you to think about cutting out 'windows'! How will this affect the strength of the package?

More Shapes and Packages

Many of the containers which we have talked about so far are 'closed' and form an almost complete outer skin, with the exception of a hole for pouring.

Of course, to protect food properly, a container must seal. But different-shaped containers present package makers with new problems. Styrene, a plastic, is often used to make butter and margarine containers. These containers show us how a very thin material can be given enough stiffness to make the container strong enough for its purpose.

Remember: When full of margarine, the container becomes a solid, and solid blocks of material resist force best.

The container with a sealed lid.

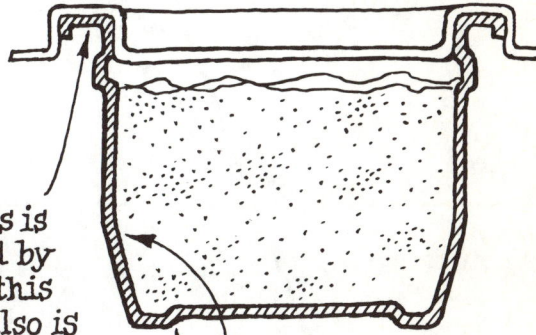

Stiffness is improved by turning this edge. It also is the seal for the container.

Shape changes here give package stiffness.

The cross-section of a margarine pack.

Developments and Nets

Look at this diagram of a package. See how the package has been planned so that there are few joins to be made. Notice that flaps and folds are included. In fact, there are many folding lines. Collect some packages and carefully unfold them to discover their shape.

These shapes are called the **developments,** or **nets.**

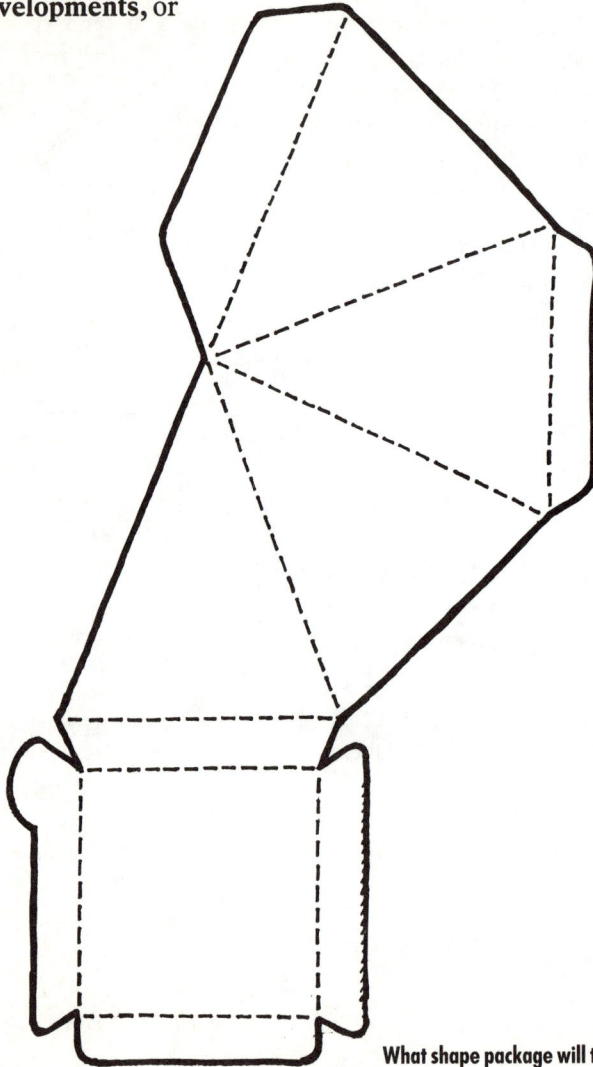

What shape package will this development fold into?

More Shapes and Packages

Whenever large areas of material are to be stiffened, features such as webs, **ribs** and **flutes** are added or surfaces are corrugated.
These sort of shapes can be found inside outer packets and provide stiffness in one direction and a type of 'crumple' zone in another.

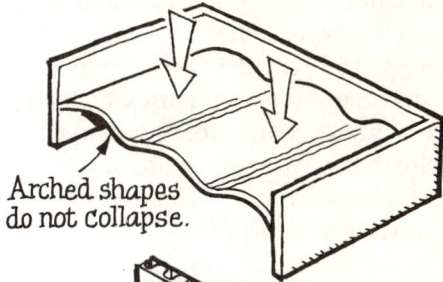
Arched shapes do not collapse.

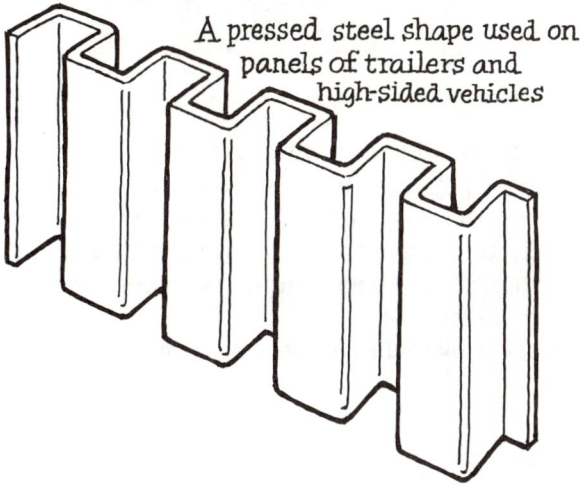
A pressed steel shape used on panels of trailers and high-sided vehicles

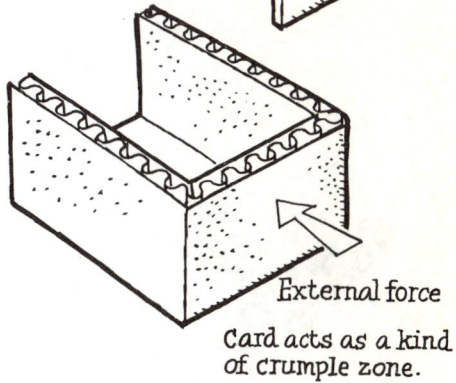
External force
Card acts as a kind of crumple zone.

Find a picture of a food tray container from an airline. Consider the requirements of such a container. It must be light in weight and cheap, because it will be thrown away after use. It must also be stiff, so that it is easy to carry. It should also be attractive, with the possibility of space for advertising the airline. Stiffness is also a feature of airlines' thin plastic cups and containers, cutlery and tin-foil containers.

Project 4

Design a range of cutlery and containers for use by an airline.
What are the important things you should think of?

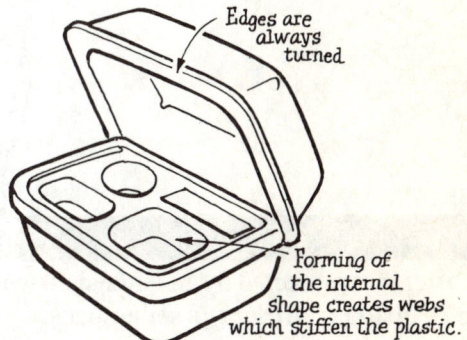
Edges are always turned
Forming of the internal shape creates webs which stiffen the plastic.

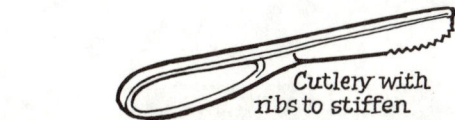
Cutlery with ribs to stiffen.

Plastic cup

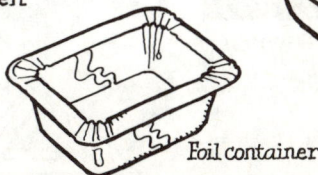
Foil container

6 Frames

Skeletons and Frameworks

We are all familiar with seeing insects and sea creatures that have a hard shell. This gives them protection and provides body support. In humans and many other animals, which are known as vertebrates, the skeleton provides protection for the body's vital organs and supports the skin and muscle. This gives a creature its own special form.

Beetle

Crab

In man-made structures, frameworks are often used. They provide the 'skeleton' on which other parts of a building are hung, and they must be capable of supporting their weight. In the roof of a house, the truss supports the tiles on the roof.
Look back to the barn structure on page 5.

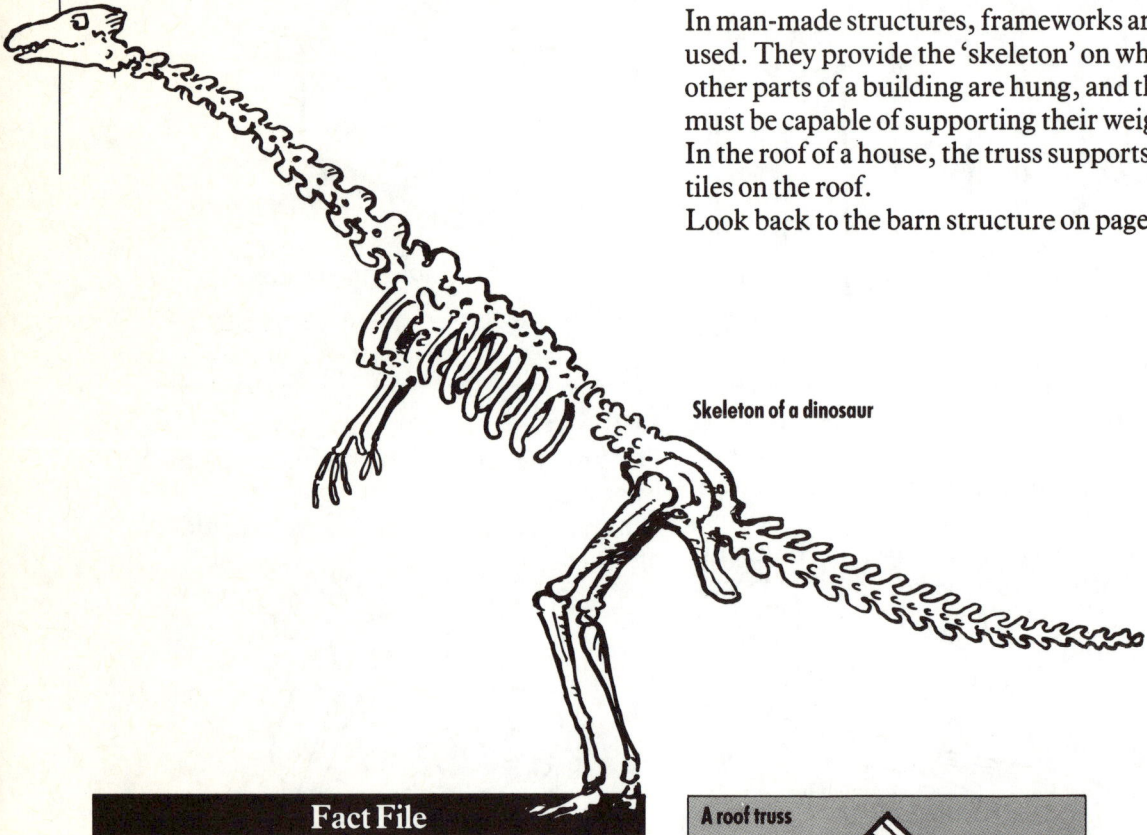
Skeleton of a dinosaur

Fact File
Lobsters are protected by an outer shell which can be up to 1 metre long in some species.

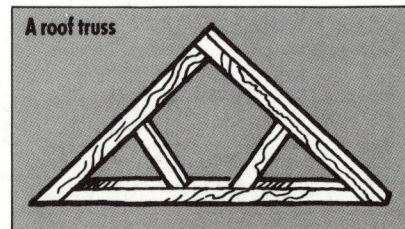
A roof truss

Tension and Ties

In a ridge tent, the frame and the guy ropes resist forces so that the material stays up. It is important that certain parts of this structure can resist compressive forces and other parts, tensile forces. Can you say what forces the labelled parts in the picture are resisting?

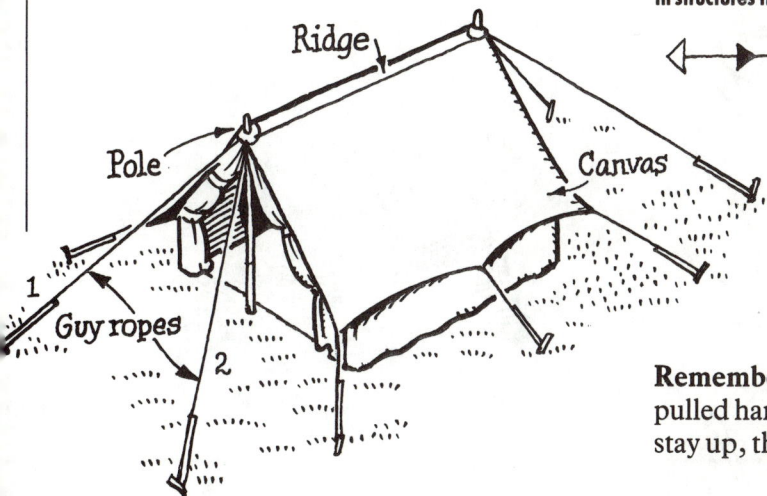

In structures those members in tension are called ties.

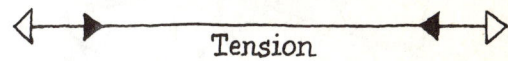

This symbol means tension.

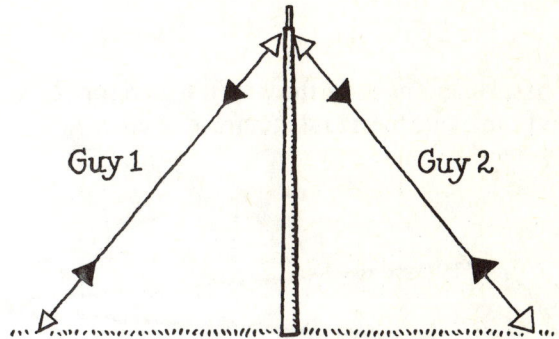

Remember: Rope will only become taut if it is pulled hard (that is, stretched). If the tent is to stay up, then guys 1 and 2 must be taut.

Therefore, they are in tension. The part of a framework which resists tensile forces is called a **tie.** You will know from your materials test that certain materials are better at resisting tensile forces than others. Ropes of fibre or steel cable are often used in structures. Look back at the diagrams of Munich Olympic Stadium on page 5. The brake cable on a bicycle is made of many strands of wire, twisted together, to resist stretching forces.

Fact File

A wire cable has been manufactured in which each strand was 28.2 cm thick. The cable was made up of 2,392 strands.

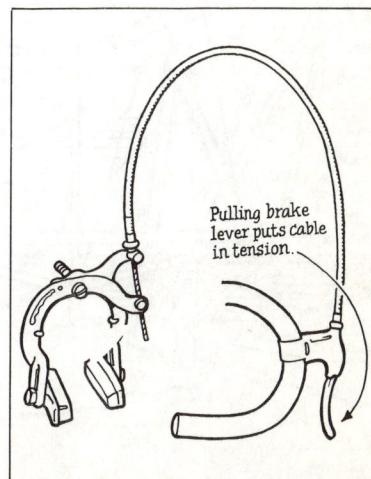

Pulling brake lever puts cable in tension.

Frames

Cables and Ropes

These diagrams show the way in which ropes
and cables are used in structural design.

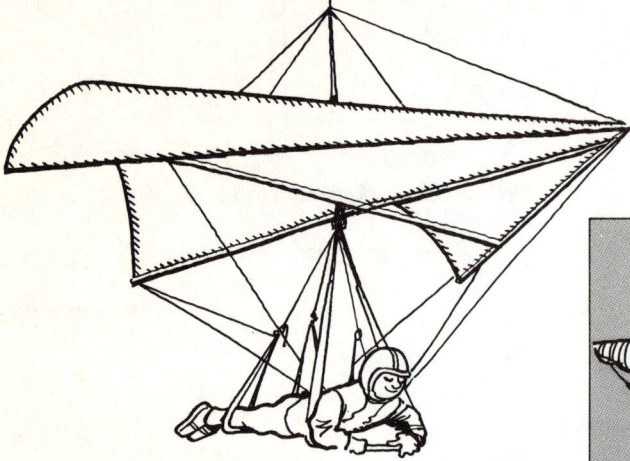

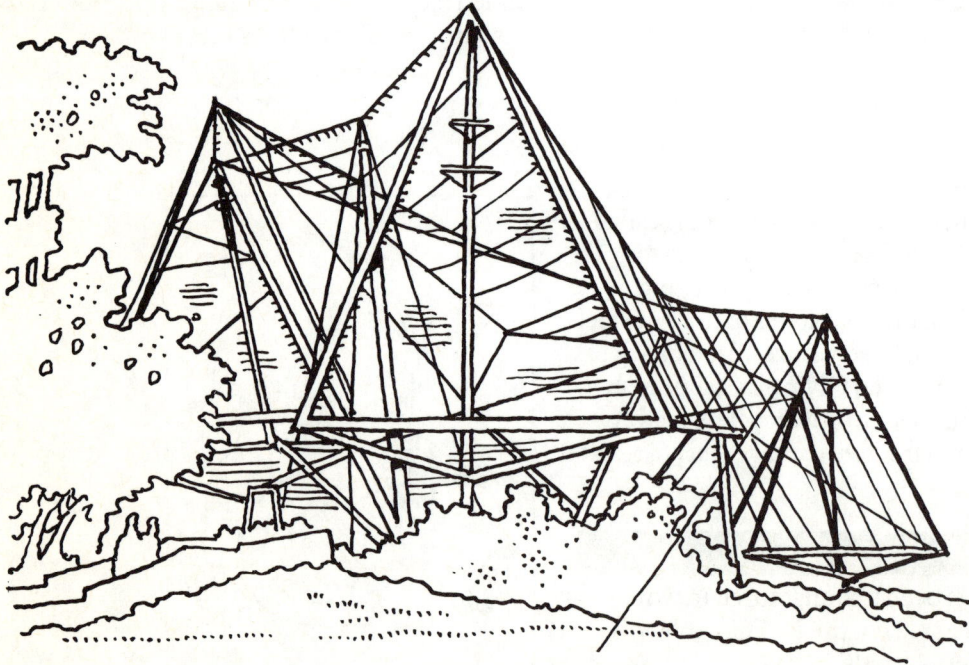

Compression and Struts

Experiment 6

Now look at the pole in the ridge tent structure. To discover the type of force acting on it, stick a piece of dowel into some Plasticine – enough for it to stand up. Attach two pieces of string, of equal length, to the dowel to act as your guy ropes.

What happens to the dowel rod if you pull with equal force on both guy ropes?

What happens if you pull hard on one side only?

What happens if you pull hard on one side only, but hold the other rope down?

Try using different materials, all of the same diameter, for the middle pole. Now use materials that have different sectional shapes. Make comments on your results, using these terms: stiffness, flex, bend and compression.

Remember: Compressive forces cause buckling and bending. The part of a framework which resists compressive forces is called a **strut**.

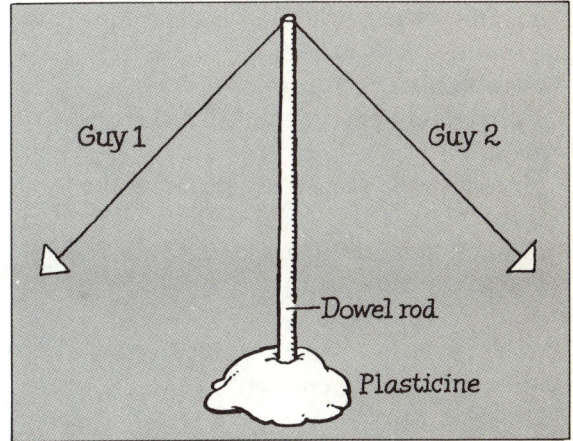

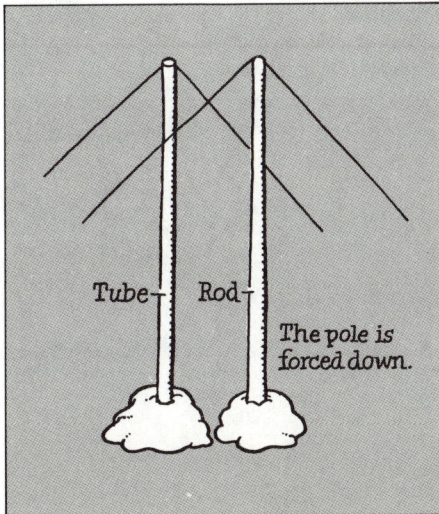

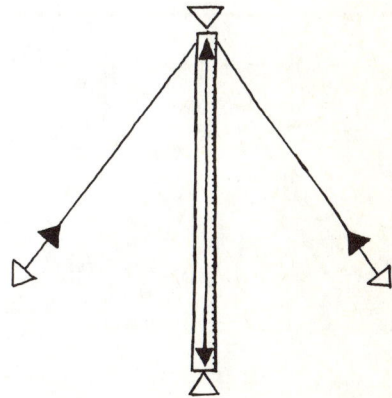

Guy 1 Guy 2
Dowel rod
Plasticine

Tube Rod
The pole is forced down.

The pole is in compression. It is a strut.

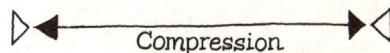

Compression

This is the symbol for compression.

Compression and Struts

Project 5: Designing a Tent

You need a tent to provide shelter each night as you hike across moorland.

Make a model of a one-person tent which is easy to put up, capable of resisting the force of wind and rain, light in weight and packs down as small as possible.

Devise tests which will help you determine how good your structure is.

Possible materials to use:

- art straws of two sizes that slip inside one another
- pipe-cleaners as a means of joining at the corners
- thin cotton sheet which can be hand-stitched or machined or glued
- thin string or thread

The diagrams show some possible starting-points and things to consider.

Art straws joined with pipe-cleaners.

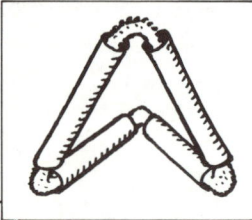

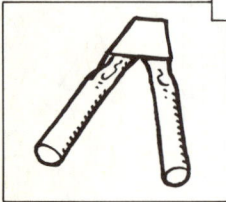

Join straws with small paper tab after squashing and gluing.

Frames are best made in triangular or arch shapes.

Traditional shape.

A different frame based on triangular shapes.

Is it easy to make the canvas shape to fit the frame?

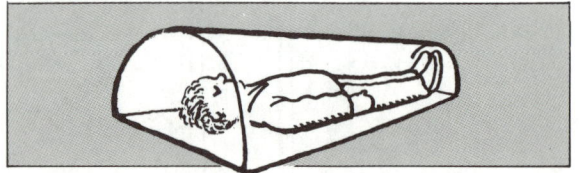

How much room is needed?

Will the frame be made of tube? What shape will the tube be?

Telescopic tubes. One straw might slide up inside another.

Forces in Frameworks

The Indian rope trick is impossible – or is it? In normal circumstances a rope could not support a person's weight. Why? The answer seems obvious, especially if you hold a piece of string up and then let it go. It is more difficult to see which forces certain parts of a wooden or steel structure are resisting.

Experiment 7

Carry out the following experiment to find out how forces act on a triangular frame, like the roof truss we looked at earlier. Make up a triangle with two lollipop sticks (or suitable lengths of wood) and a piece of string.

Rotate the triangle. As you do so, press down on the apex (top point) of the triangle.

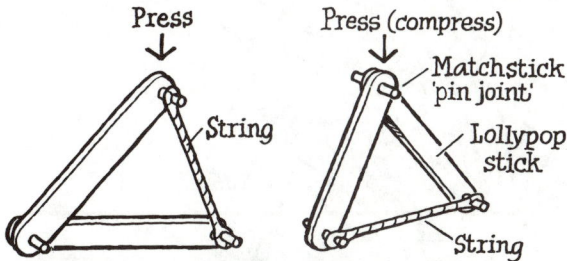

Press

String

Press (compress)

Matchstick 'pin joint'

Lollypop stick

String

When did it collapse?
When did the triangle resist the force?
Only when the string was in tension did it give support to the load.

Here are some other situations where a strut or a tie is required to stop bending and flexing.

How can you stop the bracket from bending?

How can the board be stiffened?

How can the high-diver's tower be straightened?

Triangles are important features of frame structures, even when the frame has many more sides than three.
We need to apply our knowledge of triangulation to makes things rigid.

45

Frames and Triangulation

Experiment 8

Make up the following frames from lollipop
sticks. Use matchstick pin joints to hold them
together. If you push down on these frames or load
them from one joint, they will fold. What do we
need to do to make them rigid?

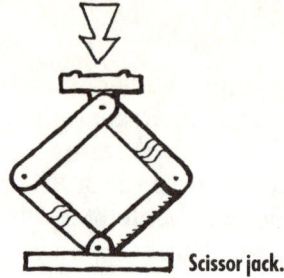

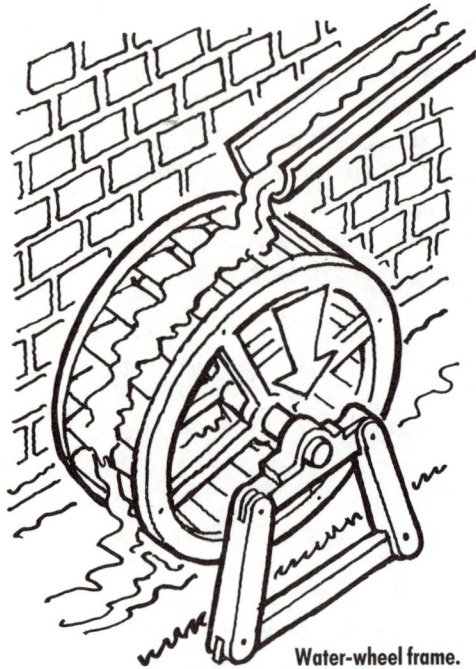

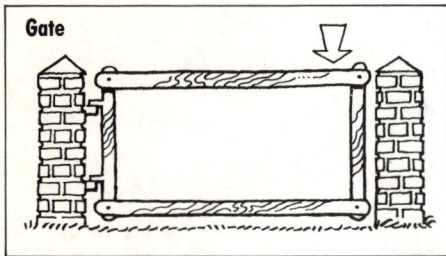

Scissor jack.

Gate

Water-wheel frame.

When designing frames, a simple rule to follow to
ensure they are rigid is:
The number of bars needed = the number of
joints × 2 − 3.
For example, the scissor jack has 4 joints:

$$4 \times 2 = 8$$
$$8 - 3 = 5 \text{ bars}$$

Continue to apply the formula to any structure
that has interlinking frames.

Solutions: All the frames have 4 joints, therefore
they should have 5 members or bars.

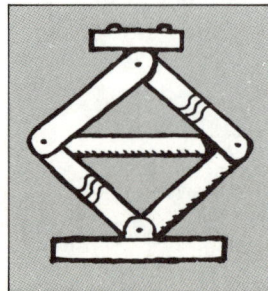

The scissor jack needs five
members to keep it rigid. Now it
will not collapse. What is special
about this extra member?

Frameworks and Triangulation

Project 6: Frameworks and triangulation

Here is a way of making frameworks with art straws.

Use this method to build a tower as high as you can which would allow a person to decorate a house or to reach to the ceiling to replace a fluorescent light.

When you scale model a solution, you need a measure of how successful you have been in your design.

First weigh your model, then measure the height you have reached. Gradually load the structure. Note exactly how it behaves, where it buckles and bends. Continue loading until structural failure occurs. A measure of success can be determined by working out this formula:

$$\text{Best model} = \frac{\text{Maximum load carried}}{\text{Weight of structure}}$$

Build the structure to scale, remembering that the more straws you use the more expensive the tower will be to buy. However, the tower should be completely safe and it should be easy to climb to the top.

Building tips: Make up the flat frames first. If any frame shape other than triangular is present, you will need to apply the formula on page 46 to ensure a rigid structure.
This means you will be using the principle of triangulation to make frames rigid.

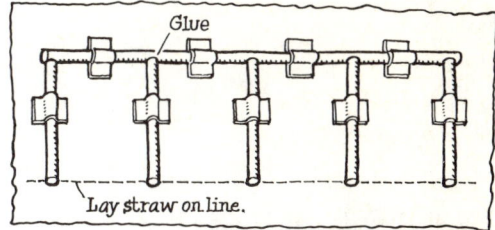

Draw the shape of the frame on grease-proof paper. Hold the straw lengths in place with masking tape. Glue at joints with PVA or hot glue gun.

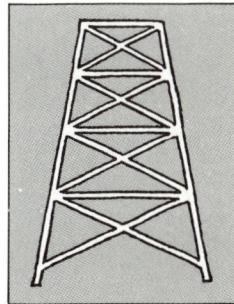

Build one frame at a time.

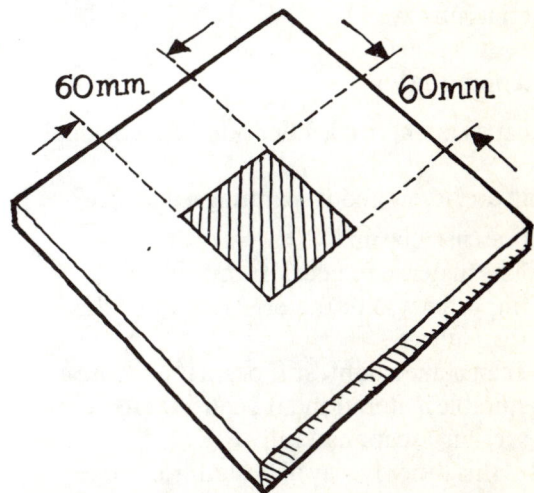

The bottom of the tower must be built within the shaded area.

Fact File

- Oil rigs in the North Sea must be built to withstand waves up to 30 metres high.
- The tallest oil platform in the North Sea is the Magnus Platform, which is 312 metres high. This is a type of tower.

7 Bridging a Wide Gap

How to Get Across

Jump! No, it's too wide.

Vault! Well, they do it in Holland.

Hang-glide. Much too energetic for me. I'm likely to fail at all three methods.

Other possible ways:

- aerial runway
- bosun's chair
- stepping on logs

Perhaps I ought to consider building a bridge!

What do I know about spanning a distance?

(a) I can use a beam.
(b) The material I make it from will be important (some materials are more flexible than others).
(c) I can make a light, stiff beam if I choose a suitable material and if I consider its cross-sectional shape carefully.

But is this the only way of building a bridge, and will the beam be the most suitable way?

Spanning Great Distances

At first you might think all bridges are one long beam that reaches from one river bank to another. Just think how long some beams would have to be if that were true. It is not always advisable to have one long beam. Can you guess why? What problems in construction would there be? Those bridges that cross motorways are often based on the simple beam, but most are changed in some way to stop deflection.

It is important that a bridge is built to withstand certain loads.
What loads have been mentioned in this book already?

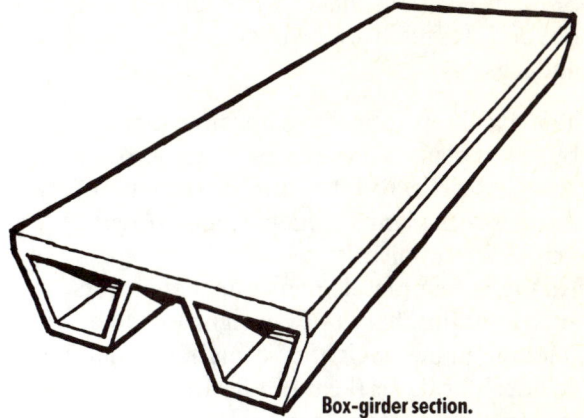

Box-girder section.

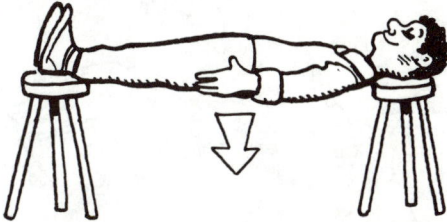

Muscle support resists body-weight.

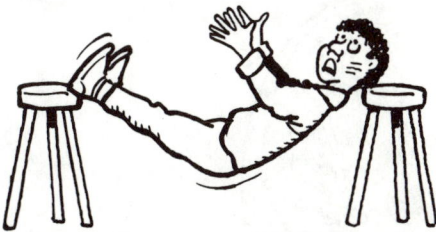

Collapsed muscles weaken.

If you were to lie across a gap between two stools, you would soon sag under your own body-weight. Your muscles would not be able to resist the pull of gravity on your body. It is easier to support your legs if they overhang a chair, but eventually they will drop to the ground.

Bridging a Wide Gap

Body-weight

Roll up a soft piece of clay or some Plasticine. Support it at both ends. Then support it only at one end. It sags in a similar way to your body and legs.

You may have noticed at some time that a builder's plank bends while resting across a stack of bricks. Builders often make temporary bridges with planks to enable them to load up skips with rubbish.

All these examples demonstrate that objects, such as beams, have **body-weight** and do bend under their own weight. We might say that the builder's bridge is flexible but sufficiently strong to support the load as it passes across it. If, however, the builder risked taking a much heavier load across the plank, it would break and structural failure would occur. The plank will not have been strong enough to support the load.

Remember: The strength of a bridge can be measured by the amount of force it takes before it fails, that is, it breaks, deforms, buckles or bends permanently.

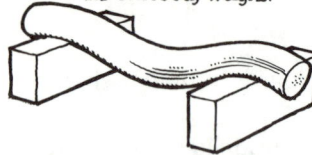
Plasticine sags under its own body-weight.

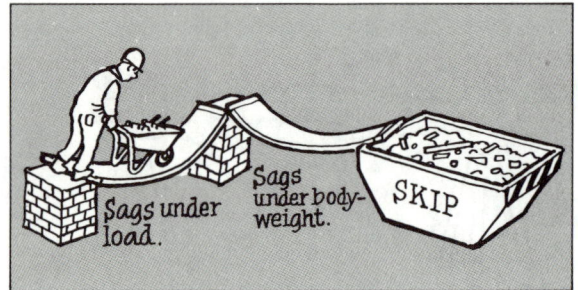
Sags under load. Sags under body-weight. SKIP

Fact File

The greatest body-weight a skeleton has to support on land is 5.7 tonnes and belongs to the African elephant.

Balance: Counter-weight

Project 7: Building a Diving Platform

You can do this with blocks of wood or bricks.

Be careful when carrying the blocks and do not carry too many at once. Imagine you have been asked to build a diving platform next to a river. Unfortunately there are rocks near the bank. How high and how far out can you build a diving platform with your bricks? It is obviously important to dive into deep water.

How far out can you go with just six bricks?

How far out can you go with ten bricks? How did you build the structure? Write your results down and record your observations. Use words like **balance** and **counter-weight** to describe your methods.

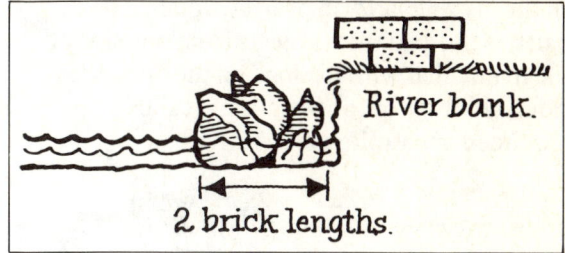

River bank.

2 brick lengths.

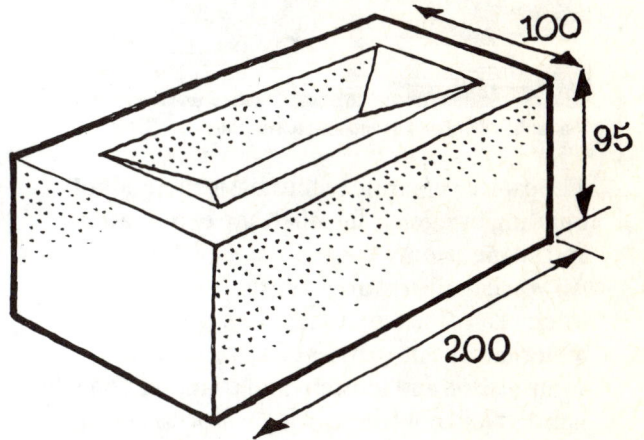

100

95

200

Not far enough.

Bridges

In the diving platform project, you had to balance the load that projects from one side of the main pillar with the load on the opposite side. This type of construction is called **balanced** or **cantilever**.

The Forth Rail Bridge is a cantilever construction.

The cantilever is one of four basic methods of building bridges. The others are **beam**, **arch** and **suspension**.

All you have learnt about structures and such things as stiffness in materials, strength in materials, the properties of materials, forces, compression and tension, and triangulation will help explain how these bridges work as structures.

Beam Bridges

On Dartmoor, large flat stones were once used to bridge the gap between river banks. These bridges are called clapper bridges. Most modern bridges, though, are made out of concrete. As you now know, concrete is a brittle material when in tension. The designer of concrete structures has to find a solution to the problem. This is done by **reinforcing,** or **pre-stressing,** the concrete beam.

The concrete is reinforced with steel bars in the region where the beam is in tension.

A reinforced concrete beam. Steel rods are placed where tension will exist in the beam when it is loaded.

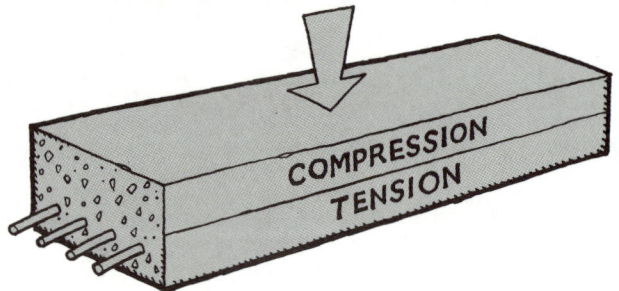

COMPRESSION
TENSION

Fact File

- The Forth Bridge is the longest cantilever bridge in Britain. Its spans are 521 metres long.
- Clapper bridges on Dartmoor and Exmoor are thought to date back to prehistoric times.

The Tarr Steps on Exmoor are an example of a clapper bridge.

Making Beams Strong

Greater strength can be given to a concrete beam by pre-stressing the steel bars before pouring on the concrete.

Because of its elasticity, the bar tries to return to its original length, thus creating a 'keeping-together' force. Longer spans can be made by using pre-stressed methods.

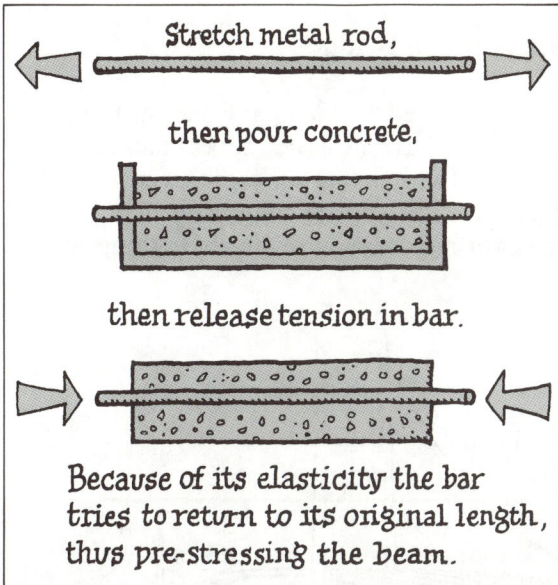

Stretch metal rod,

then pour concrete,

then release tension in bar.

Because of its elasticity the bar tries to return to its original length, thus pre-stressing the beam.

Pre-stressing a concrete beam.

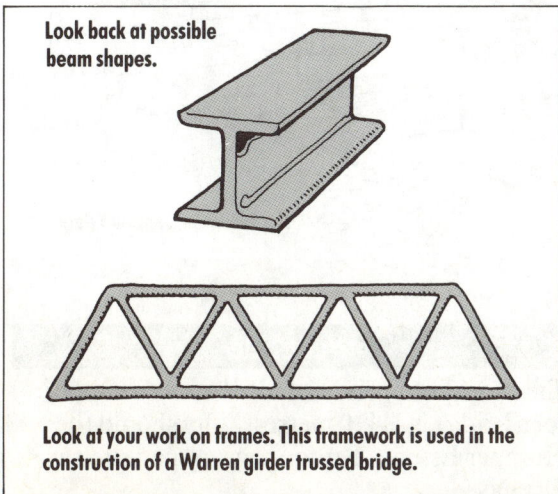

Look back at possible beam shapes.

Look at your work on frames. This framework is used in the construction of a Warren girder trussed bridge.

Beams can also be made out of wood or steel. When made of steel, the beam is also given a particular shape. You should be able to explain why. Think back to your beam making.

The large amount of material at the top and bottom, and the distance between it, enables it to be much stiffer because the internal forces set up in the beam when under load are large. Some bridges gain greater stiffness by having trusses, which are types of framework along their edge. Beams of various sorts may be built into bridges.

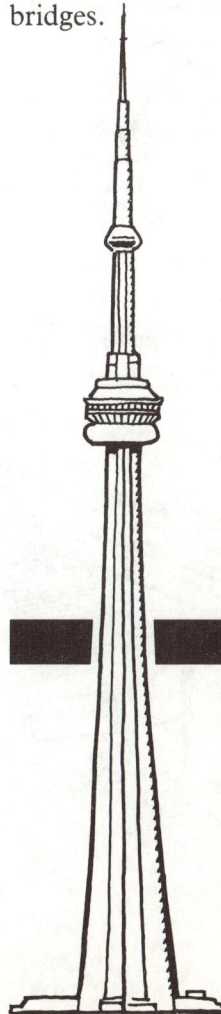

Fact File

The CN Tower in Toronto, Canada, is 555.33 metres tall. It is made of reinforced as well as pre-stressed concrete.

Bridging a Wide Gap

Bridges

Arch Bridges

To understand how the arch works, take a thin piece of card or thin-section wood. In its flat form it bends when loaded. Hold it in an arch shape. It is unable to deform and the force of the load runs to the ground. It is very important to have solid support at points A and B.

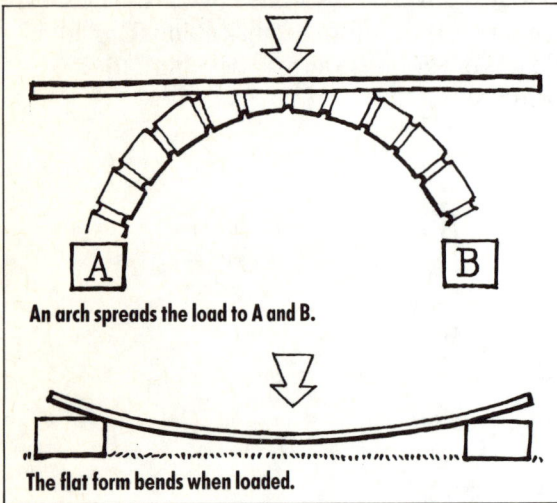

An arch spreads the load to A and B.

The flat form bends when loaded.

Suspension Bridges

Cables are used to hold up the road surface. They often have an arched appearance and are capable of supporting surfaces that span great distances. They must have great tensile strength and be firmly anchored to the ground.

The Severn Suspension Bridge.

Fact File

The world's longest distance between supports for a bridge is 1,410 metres. Why should the Humber Estuary Bridge have to span such a distance?

Bridges

Project 8: Building a Bridge

Here is a list of materials you can use in your bridge design. They are grouped together, but they might be combined to enable different constructions to be made. Your teacher will limit you to the types and amounts of material you can use.

Group A Art straws 1 piece of card Thread	**Group D** Strips of wood Jelutong Matchsticks 1 piece of card
Group B Jumbo straws and card only	**Group E** Uncooked spaghetti PVA glue Paper Card
Group C Uncooked spaghetti Glue gun 1 piece of card	**Group F** A long length of card which can be cut and straws or jelutong

Work in groups and imagine that you are a company competing against another to win a bridge-building contract. Choose one of the suggested situations at the end of this section from which to work. The overall cost of materials and ease of construction will be very important factors in choosing the best-designed bridge, but other factors must also be considered. Will the bridge blend in with the natural surroundings? Will it stand up to all those forces it is likely to be subjected to?

Bridge Problems

Wood strips

Paper triangles

Hints on Construction

- Make up frames or flat shapes first.
- Frames with straws. Draw a pattern on some grease-proof paper first. Cut each straw to length and tape it down in position. Use PVA glue or a hot glue gun to join the straws.
- Spaghetti can be joined in the same manner.

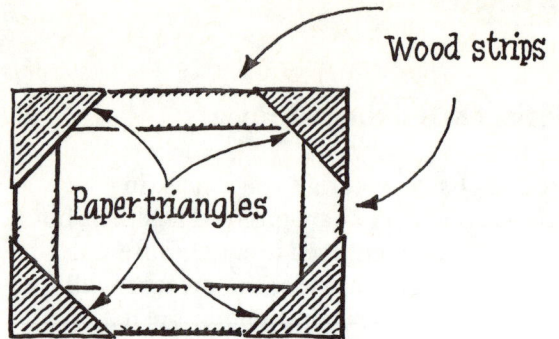

Pin joint

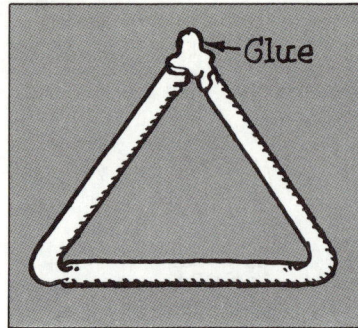

Glue

- Frames with wood. A template should also be used to determine the arrangements of the various parts. Flat frames can be manufactured with paper triangles.
- Card can be glued to wood or straws or held to certain shapes between frames. Thread can be tied to different members or, better still, threaded through small drilled holes in wood or pierced holes in straws.

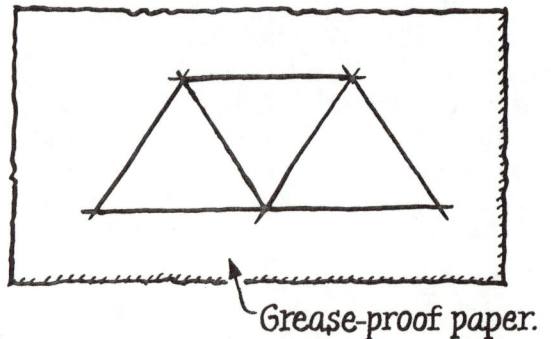

Grease-proof paper.

Always make a template. Draw out the shape of the framework, then lay pieces on top.

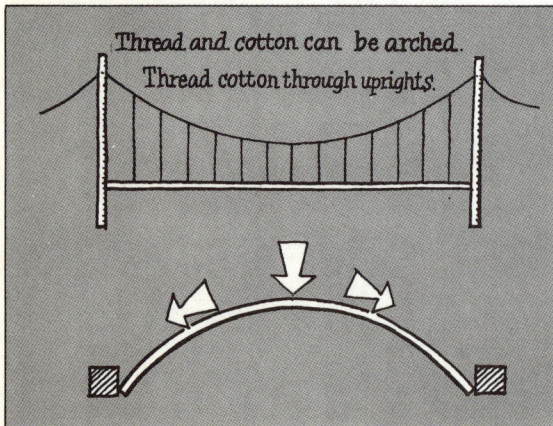

Thread and cotton can be arched.
Thread cotton through uprights.

Try out some of your own methods of joining. These are only suggestions.

Bridge Problems

Situations

1 A rocky gorge in a national park. Walkers want to cross.
The bridge should be built over a gap to the scale of 10 mm to the metre.

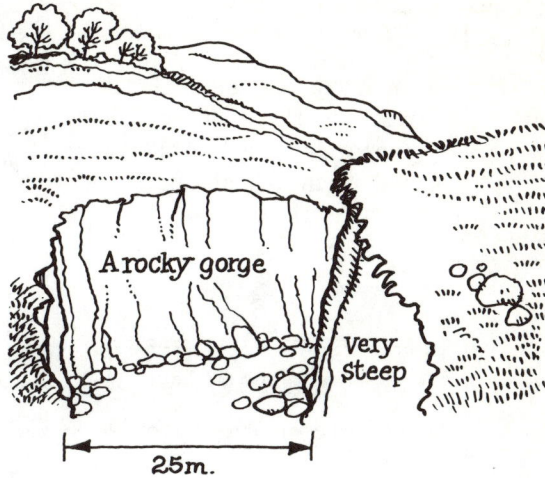

A rocky gorge

very steep

25 m.

2 A bridge over a road. Other traffic must pass over the road. No obstruction is allowed on the road below. There are high earth banks each side.
Build to a scale of 10 mm to the metre.

3 A bridge over a wide river with a solid rock bed; to carry a railway.
Build to a scale of 1 mm to the metre.

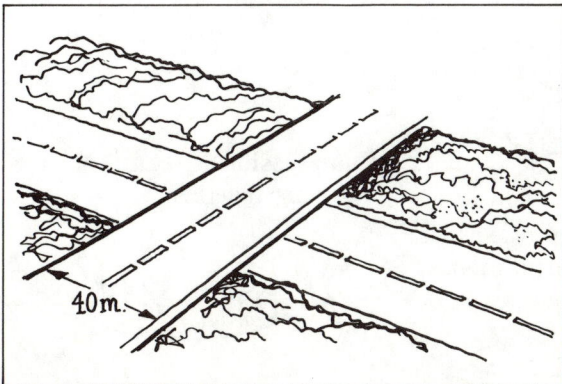

40 m.

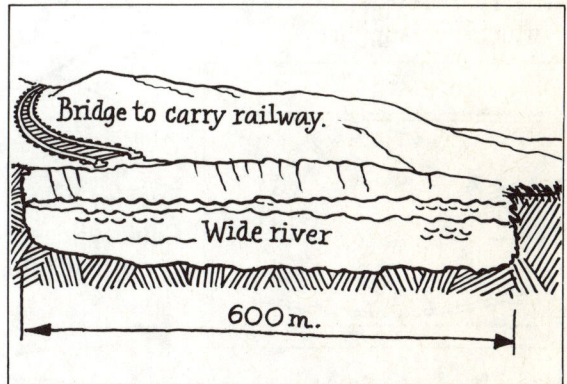

Bridge to carry railway.

Wide river

600 m.

Mini-Dictionary

Acrylic A type of plastic.

Apex The tip or topmost point of a triangle.

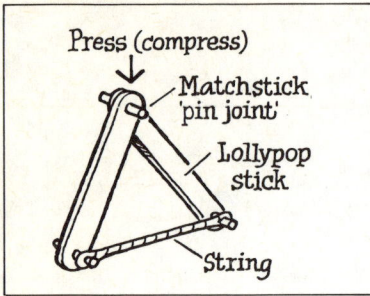

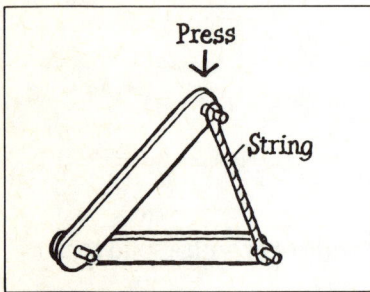

Press (compress)
Matchstick 'pin joint'
Lollypop stick
String

Press
String

Appearance How good something looks.

Arch A curved part of a structure which gives support.

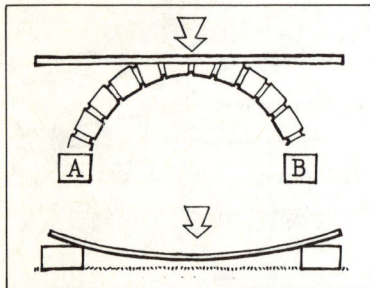

A   B

Axle A bar connecting wheels.

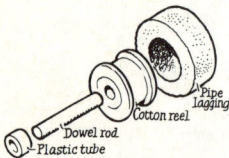

Pipe lagging
Cotton reel
Dowel rod
Plastic tube

Balance Two sides being equal.

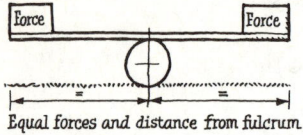

Force   Force

Equal forces and distance from fulcrum.

Beam A length of material that reaches across or out, which is supported at one end at least, and which is capable of carrying a load.

Wooden beam. One person, no deflection.

Bend Force out of a straight shape.

bends
STEEL

Body-weight A structure's own weight.

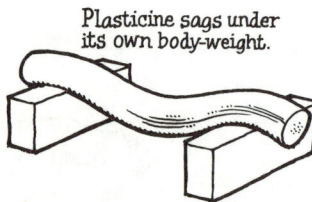

Plasticine sags under its own body-weight.

Buckled Folded or crushed.

Cable A line of rope or steel that can hold things in position.

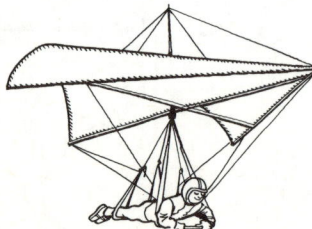

Cantilever A beam supported at one end.

Chassis A frame on which things are fixed.

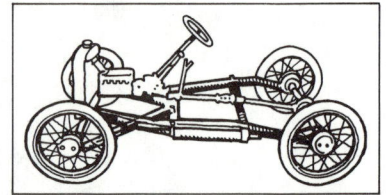

Collapse A change in shape or position, like falling down or buckling.

Compression Squeezing or forcing together.

PUSH   COMPRESSION   PUSH
INTERNAL FORCE

Constraint In design, this means a condition that must be met.

58

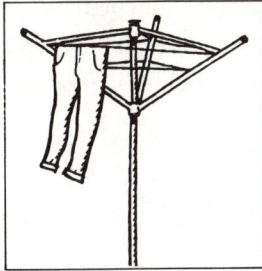

Construct Put together.

Contain Hold together or within.

Counter-weight A mass placed on the opposite side of a pivot to maintain balance.

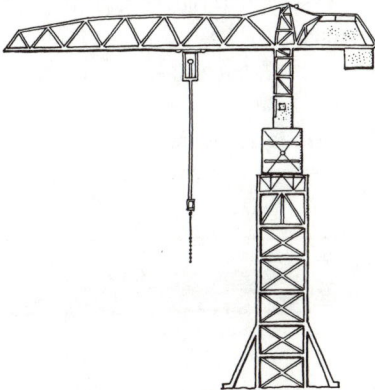

Cross-section The end shape of material when cut through.

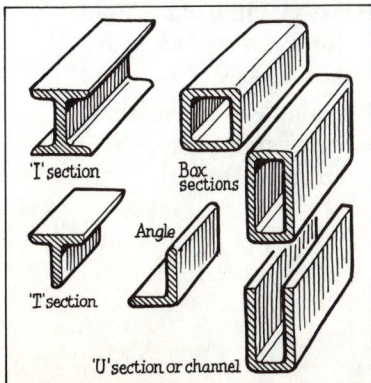

'I' section

Box sections

'T' section

Angle

'U' section or channel

Crumple zone An area which collapses when subjected to force.

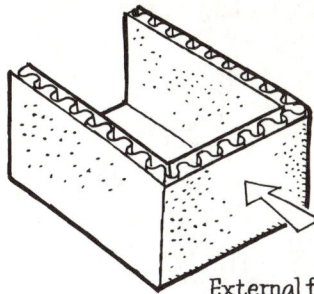

External force

Card acts as a kind of crumple zone.

Deflect Move as a result of force.

deflection

One diver load

Destructive testing Testing until failure occurs.

Development The shape of a package when opened out and laid flat.

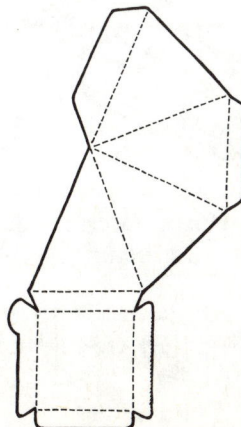

Dynamic force A moving, destructive force.

Dynamic

Elasticity A material's ability to return to its original shape after being deformed.

Equilibrium A state of balance.

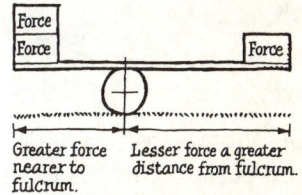

Force
Force
Force

Greater force nearer to fulcrum.

Lesser force a greater distance from fulcrum.

External force An outside force acting on a structure.

EXTERNAL FORCE GREATER THAN

PUSH   INTERNAL   FORCE   PUSH

THEN THE STRUCTURE BUCKLES

Failure In a structure, this is when a member breaks, fractures, buckles, bends, flexes, twists or stretches and is permanently distorted.

Flexibility The ability to bend and move without distortion or damage.

Flute A groove with a curved bottom.

Force An influence which causes or resists motion (movement).

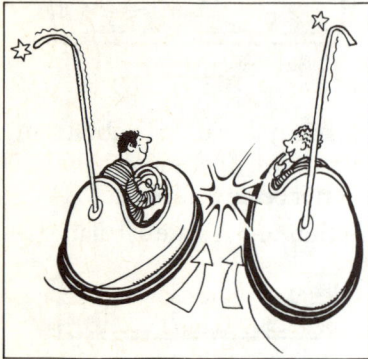

Frame/Framework The basic parts of a structure when put together.

Function What something has to do.

Graph A plotted record of test results.

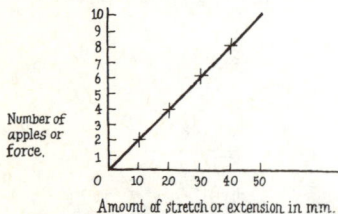

Number of apples or force.

Amount of stretch or extension in mm.

Gravity A material's attraction to the centre of the earth. The attraction of a smaller mass towards a larger mass.

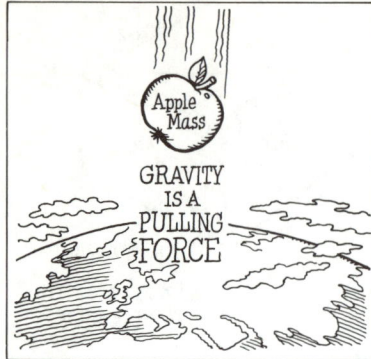

Apple Mass

GRAVITY IS A PULLING FORCE

Guy Rope that holds up a tent.

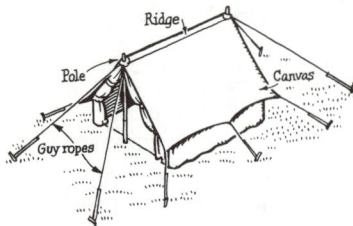

Ridge

Pole

Canvas

Guy ropes

Hardness The ability to resist scratching or being dented.

Hexagonal Six-sided.

Honeycomb

Internal force Forces inside the parts of a structure.

TENSION

PULL

PULL

INTERNAL FORCE

Kilogram A unit of mass.

Load An amount of material or the result of a force.

MAX LOAD

STOP! NO MORE!

Mass The amount of substance in something.

Model A smaller or larger version of a structure, but made to the same proportions.

Moment A turning-force acting on a beam.

Monocoque construction A single-skin structure.

Net The shape of a flat material which, when folded, produces a package or container.

Newton A unit of force. All forces are measured in newtons (N).

Pin joint A simple method of connecting parts of a structural model.

Pin joint

Pre-stress Set up internal forces in steel by stretching it.

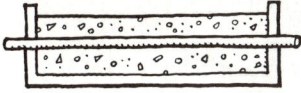

PVA Polyvinyl acetate emulsion, one type of glue.

Protect Keep safe.

Pylon Tall support.

Reinforced Strengthened.

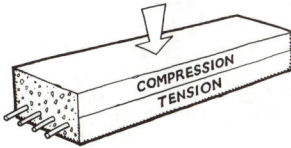

Ribs Shapes that stiffen material.

Ridge The line of the join where two surfaces meet.

Ridge tent A fabric shelter where two flat surfaces meet to make a roof.

Rigid Not flexible — something that cannot be bent.

Score Partially cut through.

Span Reach across or over.

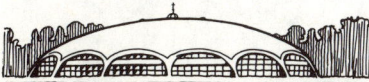

Spar Supporting pole.

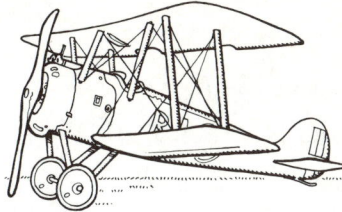

Static force A force at rest; still.

Sterilize To make germ free.

Stiffness The ability of a structure to withstand being bent, twisted, stretched or changed in shape.

Strain $\dfrac{\text{Change in length}}{\text{Original length}}$

Strength Measured by the amount of force a structure takes before failure or material before it breaks.

Stress $\dfrac{\text{Force}}{\text{Area}}$

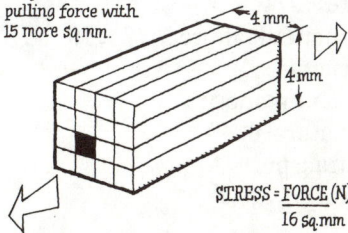

1 sq. mm shares the pulling force with 15 more sq. mm.

$$\text{STRESS} = \dfrac{\text{FORCE (N)}}{16 \text{ sq. mm}}$$

Strut A member of a framework which resists compressive forces.

Styrene A type of plastic.

Suspend Hang up or onto.

Taut Pulled tight.

Template A shape which can be copied.

Tension Pulling or stretching forces.

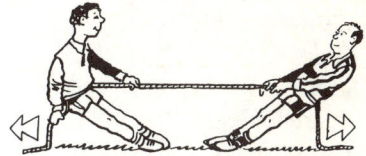

Tie A member of a framework which resists being stretched.

Toughness The amount of energy a material can absorb before breaking.

Transport Ways of carrying people and goods about.

Triangulation The use of triangles in frameworks.

Truss A timber support in the roof of a house.

Weight A force, measured in newtons (N). It is to do with the attraction of a mass to the earth through the force of gravity.

Index